A Treatise on the Science of Music

D ANIEL M.G.S. R EEVES

CAMBRIDGE
UNIVERSITY PRESS

CAMBRIDGE UNIVERSITY PRESS

Cambridge, New York, Melbourne, Madrid, Cape Town,
Singapore, São Paolo, Delhi, Tokyo, Mexico City

Published in the United States of America by Cambridge University Press, New York

www.cambridge.org
Information on this title: www.cambridge.org/9781108038805

© in this compilation Cambridge University Press 2011

This edition first published 1853
This digitally printed version 2011

ISBN 978-1-108-03880-5 Paperback

CAMBRIDGE LIBRARY COLLECTION

Books of enduring scholarly value

Music

The systematic academic study of music gave rise to works of description, analysis and criticism, by composers and performers, philosophers and anthropologists, historians and teachers, and by a new kind of scholar - the musicologist. This series makes available a range of significant works encompassing all aspects of the developing discipline.

A Treatise on the Science of Music

First published by the house of Novello in 1853, and later reprinted, this was one of the earliest treatises to take a scientific as well as a practical approach to the discussion of music. Written before Wagner had begun work on *Tristan*, this work can be seen as a response to the growing interest from the amateur in the 'science' of music. Little is known about the author, Daniel Reeves, who declares that that 'the idea of music comprises both an art and a science: the art consisting in the power of performing ... ; the science, in an acquaintance with the system on which the constituent sounds ... depend'. Using numerous examples, Reeves explains the basics of musical notation, and includes a lengthy mathematical analysis of the ratios of tones and intervals, underlining his belief that an understanding of music should be 'a necessary branch of every gentleman's education'.

Cambridge University Press has long been a pioneer in the reissuing of out-of-print titles from its own backlist, producing digital reprints of books that are still sought after by scholars and students but could not be reprinted economically using traditional technology. The Cambridge Library Collection extends this activity to a wider range of books which are still of importance to researchers and professionals, either for the source material they contain, or as landmarks in the history of their academic discipline.

Drawing from the world-renowned collections in the Cambridge University Library, and guided by the advice of experts in each subject area, Cambridge University Press is using state-of-the-art scanning machines in its own Printing House to capture the content of each book selected for inclusion. The files are processed to give a consistently clear, crisp image, and the books finished to the high quality standard for which the Press is recognised around the world. The latest print-on-demand technology ensures that the books will remain available indefinitely, and that orders for single or multiple copies can quickly be supplied.

The Cambridge Library Collection will bring back to life books of enduring scholarly value (including out-of-copyright works originally issued by other publishers) across a wide range of disciplines in the humanities and social sciences and in science and technology.

A TREATISE

ON

THE SCIENCE OF MUSIC,

BY

DANIEL M. G. S. REEVES.

" Untwisting all the chains that tie
The hidden soul of Harmony."
MILTON.

LONDON :

J. ALFRED NOVELLO,
69 DEAN STREET, SOHO, AND 24 POULTRY;
AND IN
NEW YORK, AT 389 BROADWAY.

1853.

CORRIGENDA.

At p. 17, line 20, instead of the words " or *of* F major," read " or F major."

At p. 36, at foot, at the words " See Appendix, Note (W)," insert the mark of reference ; thus, " † See Appendix, Note (W)."

At p. 86, affix to the Chant the name of the composer, "Heathcote."

At p. 111, line 18, for " concurrence," read " occurrence."

At p. 112, at foot, instead of " See Appendix, Note (JJ)," read " Vide the Synopsis, sup. p. 16."

At p. 158, line 2, dele " p. 46."

At p. 205, line 28, instead of " A, B♭, C, D, E, F, G," read " G, A, B♭, C, D, E, F, G."

At p. 213, line 20, instead of the words " with the 8th B omitted," read " with the 8th B̄ omitted."

At p. 215, line 7, instead of the words " or into that of a different tonic," read " or into the minor key, not relative, of a different tonic."

At p. 220, instead of " D♯, F, A, C," read " D♯, F♯, A, C."

PREFACE.

THE idea of Music comprises both an art and a
science; the art consisting in the power of performing
an air or piece; the science, in an acquaintance with
the system on which the constituent sounds of the air
or piece, and the successions and combinations of those
sounds, depend. The science is essential to the complete efficiency of the artist, unless he means always
to confine his performance to the compositions of others,
without aspiring to be himself a composer. Nor can
there be anything unreasonable in considering it even
as a necessary branch of every gentleman's education,
whether he intends to be conversant with the art or not;
for it is indispensable to the power of fully appreciating
the works of the great masters, and to the formation
of a critical judgment and a true taste ; and it involves,
besides, when thoroughly investigated, some curious
facts connected with the philosophy of sound. And
yet, however common the art, an acquaintance with

the science is comparatively rare, except in that limited and imperfect sense which is implied in a mechanical knowledge of the scale, the keys, and the time table. This is attributable in some degree to the dryness and the difficulty by which the study of it is found to be attended. In the present work, which is intended for the benefit of those who apply themselves to that study, the author has attempted to relieve the dryness by the introduction of various examples from eminent composers*, and of some occasional disquisitions of mere curiosity; and to diminish the difficulty by an invariable attention to clearness and precision in his manner of treating the subject.

Those already acquainted with music will perceive a peculiarity in his method of *figuring* chords, on which it may be right to offer some explanation in this place. He departs from the custom usually observed, where the interval of a third occurs, introductory of a sharp, flat, or natural, foreign to the key, of putting a mark of ♯, ♭, or ♮, under the figuring; and uniformly omits these marks†. He does this, because he conceives them to be really called for only in the case of a mere figured *bass*. When the chords are given full (as is always the case in this work), they appear

---

* Most of the examples bear the composer's name. Of those that are anonymous, the authors are unknown; but they have been selected, in some instances, from Mr. Shield's work on Harmony.

* Two exceptions, however, will be found at p. 64, No. 33, No. 35, where a 3rd with a ♯ is so marked in the figuring. They escaped observation during the revision for the press.

to him to be wholly useless; the introduction of the
sharp, flat, and natural, into the chord being as immediately apparent on sight of the chord, as it can be upon
the figuring.

It would be a further improvement, however, in
the figuring of chords, if it were done with such
marks, in addition to the figures, as to show at a glance
whether any given interval in them be major or minor—
belonging to the scale, or extraneous. For this purpose
the figures might be marked as follows:

A semitone	·2
A second	2
A minor 3rd	·3
A major 3rd	3
A 4th	4
A tritone	4·
A minor 5th	·5
A 5th	5
A minor 6th	·6
A major 6th	6
A minor 7th	·7
A major 7th	7
An 8th	8
A minor 9th	·9
A major 9th	9
A diminished 4th	··4
A diminished 5th	··5
A diminished minor 6th	··6
A diminished minor 7th	··7
An augmented 2nd	2̶
An augmented 4th	4̶
An augmented 5th	5̶
An augmented major 6th	6̶

The effect of such a system of figuring may be
sufficiently illustrated by the following short example* :

And the advantage that would result from it is
sufficiently apparent.   But so great an innovation was
more than the author had courage to introduce into
the work.

---

* See the same example figured in the common way, post, p. 153.

CONTENTS.

CORRIGENDA.

At p. 17, line 20, instead of the words " or *of* F major," read " or F major."

At p. 26, Example of Fourteenth minor, read

At p. 27, Example of Seventeenth major, read

At p. 33, line 12, instead of " E, G, C, E, and G, C̄, E, G," read " E, G, C, Ē, and G ,C, E, Ḡ."

At p. 36, at foot, at the words " See Appendix, Note (W)," insert the mark of reference ; thus, " † See Appendix, Note (W)."

At p. 86, affix to the Chant the name of the composer, " Heathcote."

At p. 111, line 3 from the bottom, instead of " this mode," read " the minor mode."

At p. 111, line 18, for " concurrence," read " occurrence."

At p. 112, at foot, instead of " See Appendix, Note (JJ)," read " Vide the Synopsis, sup. p. 16."

At p. 113, line 7 from the bottom, instead of " when," read " where."

At p. 136, line 7, after the words " when placed on the tonic," insert the words " or dominant."

At p. 147, last chord but one, transfer the ♮ from the B to the D.

At p. 170, in the foot note, instead of " 109," read " 100."

At p. 205, line 28, instead of " A, B♭, C, D, E, F, G," read " G, A, B♭, C, D, E, F, G."

At p. 212, line 3 from the bottom, instead of " Holder," read " Holden."

At p. 213, line 20, instead of the words " with the 8th B omitted," read " with the 8th B̄ omitted."

At p. 215, line 7, instead of the words " or into that of a different tonic," read " or into the minor key, not relative, of a different tonic."

At p. 220, instead of " D♯, F, A, C," read " D♯, F♯, A, C."

CHAPTER I.

A note is any musical sound ; that is, any sound giving pleasure to the ear.

The ear distinguishes notes from each other in various respects, and, among others, in respect of *height ;* a difference which has been ascertained by experiment to depend upon the relative number of vibrations made in a given time by each of the sounding bodies ; a body whose vibrations are more frequent, emitting a higher note than one whose vibrations are less frequent.

The degree of the height of a note is called its *pitch ;* and the amount of the difference of pitch between any two notes is called the *interval* between them.

The *scale* is a series of notes, each successively ascending or descending in pitch, at certain prescribed intervals from each other.

The number of notes in the scale is eight ; and they are denominated (counting upwards) as the 1st, 2nd, 3rd, 4th, 5th, 6th, 7th, and 8th ; the 8th being also called the *octave* of the 1st.

B

Any note or musical sound whatever may be made the first of a series; so that the scale determines nothing as to the particular notes comprised in it; it only determines their number and the intervals at which they stand in respect to each other.

The 1st and 8th notes of the scale are in such relation to each other, that, though differing considerably in pitch, they are nevertheless felt by the ear to be very similar, or in a manner identical; a circumstance owing to the simplicity of the ratio (viz. 1 : 2) between the numbers of the vibrations of the strings, or other sounding bodies, by which these notes are respectively emitted.

The eight notes of the scale collectively are said to be the *key* of the 1st note, and that note is consequently called the *key note*. Thus, if the scale begins with the note known among musicians as C, the eight notes of which this is the 1st are called the key of C, and C is called the 1st of the key or (more usually) the *key note*.

The intervals comprised in the scale, as between *contiguous* notes, are of two kinds—the one greater, called a *tone* (or a *second*), the other less, called a *semitone ;* some notes having a tone between them, and some a semitone.

The scale admits of two different arrangements, as to the order of succession of the tones and semitones ; one arrangement being called the *major mode,* and the other the *minor mode.*

The first arrangement, that is, the major mode, is as follows :

(Ascending)—Tone, tone, semitone, tone, tone, tone, semitone.

(Descending)—The same reversed.

The second arrangement, that is, the minor mode, is as follows :

(Descending)—Tone, tone, semitone, tone, tone, semitone, tone.

(Ascending)—The same reversed, subject to variations, to be afterwards explained.

From what has been said, it may be collected that the system admits, in either mode, a variety of keys; because any note whatever may form the first of the series, and, with the seven notes that follow, will consequently constitute a key. It remains, however, to explain more particularly the manner in which the several keys are produced.

And, first, in the major mode.

Some particular sound being first fixed*, and a name (suppose C) assigned to it, it is then made to form the 1st or key note of the key of C; the notes of which are designated (ascending) C, D, E, F, G, A, B, $\overline{\text{C}}$; for the 8th (or octave of the 1st) being in a manner the same (as already explained) with the 1st, is denominated by the same letter. And as the upper C may in turn be made the 1st of a similar series ascending, and the lower (or original) C the 1st note of a similar series descending (every note in each new series being tuned an octave to, or replicate of, the note of the same denomination in the first or original series), and as this progression may be pursued to any extent, until the sounds become too high or too low to be musical†, we have thus a considerable number of sets, or repetitions (or different *octaves*, as they are called) of the key of C, differing from each other only in pitch.

This key of C (in the major mode), with its intervals of tones and semitones marked below the letters, will be as follows:

C	D	E	F	G	A	B	$\overline{\text{C}}$‡
| tone |	tone	|semi|	tone	| tone |	tone	|semi|

and this expresses, not only the 1st or original octave, but all the others.

But here it becomes necessary, at the expense of some

---

* See Appendix, Note (A).

† See Appendix, Note (B).

‡ A dash over a letter, as here, will be used in this work to express a higher octave. Thus, $\overline{\text{C}}$, means the note that is an octave above C.

digression, to explain the manner of designating notes, or, as it is called, the subject of *notation*.

The letters C, D, E, F, G, A, B, C, constitute (as we have seen) the *names* of the notes; but are themselves expressed to the eye, not usually by their alphabetical characters, but by other characters placed on *staves* (as they are ealled), that is, on groups of parallel lines, and on the spaces intervening between them.

These characters are of different forms, for the purpose of expressing their relative value to each other as regards *time*, or the duration of the sound; and are, with their relative values, set forth in the following Table:

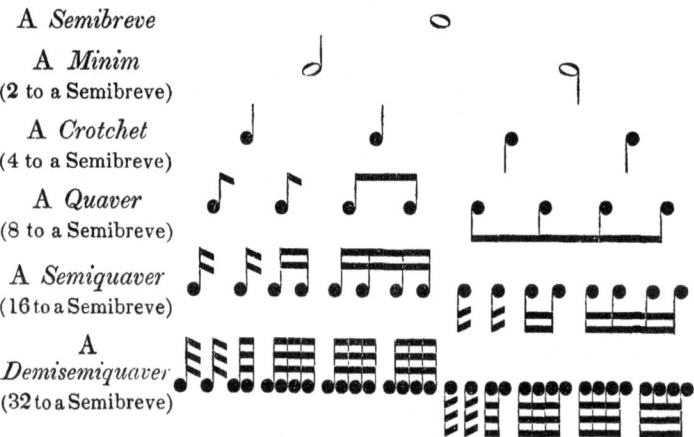

A *Semibreve*

A *Minim*
(2 to a Semibreve)

A *Crotchet*
(4 to a Semibreve)

A *Quaver*
(8 to a Semibreve)

A *Semiquaver*
(16 to a Semibreve)

A
Demisemiquaver
(32 to a Semibreve)

The number of lines in the staves, on which these characters are written, are five; but other short ones, called *ledger lines*, are added above and below the main body of the staff, as occasion requires. The staves bear also certain marks called *Clefs*, of which two only are now popularly used, and are as follows: 𝄞 and 𝄢 : ; the first being called the G or treble clef, the second the F or bass clef. The treble clef is placed on the second line of the staff, counting upwards, and signifies that the note G (at a certain pitch, called by musicians *treble*) is to be placed on that line. The bass clef is placed on the fourth line of the staff, counting upwards, and signifies that the note F (at the pitch called by musicians *bass*, and which is lower than the treble) is to be placed on that line. The place of one particular note being thus fixed in each of the staves on which these clefs are put, determines of course the place of all the remaining ones; and the two clefs each designate a series (one at treble, and the other at bass pitch, of the same notes, C, D, E, F, G, A, B, C, each of which has its own place in the staff to which that particular clef is attached. All this will be more clearly understood by the following representation of two staves, one in the bass, and the other in the treble clef, with the characters of the notes (in this case supposed to be *crotchets*) in our original key of C, placed in due order upon the lines and spaces, and the alphabetical letters subjoined to each note, for the purpose of clearer explanation

It is to be observed, that in the first staff the upper C is the same note as the lower C in the second staff, and at the same pitch. The two staves together, therefore, give more than three consecutive series of octaves, and by increasing the number of ledger lines (of which only two are above used in each staff, viz. one at the first note, and another at the last), the pitch in each staff may be yet further extended upwards and downwards. It is found, in practice, that these two clefs (without more), extended as occasion may require by ledger lines, are in general sufficient. A greater variety of denomination, indeed, than bass and treble, is used to express the difference of pitch ; four principal degrees of elevation being recognised in reference to this subject, viz. *treble, alto, tenor,* and *bass,* which lie in that order (descending) with respect to each other ; and each of these parts (besides several others) had formerly always its separate clef* ; but to avoid so inconvenient a variety of notation, the treble and bass clefs are now used almost exclusively, and it is in the range of the connected series of notes which they together comprise, that all sounds, of whatever pitch, find their place†.

To return to the subject of keys ; our consideration of which for the present has been confined to the key of C. We are next to explain how other keys are added to the system. To effect this, it is obvious to proceed in the first place by making *others* of the original notes, viz. D, E, F, G, A, B, successively form the key note, instead of C. They are not taken, however, in that order; the note first employed for the purpose being the fifth note ascending from C, viz. G. Taking G then as the new key note, the series of the new key, supposing all the old notes to be retained, would be as follows, viz. G, A, B, C, D, E, F, G̅‡. But such a series would be incorrect, not being in conformity with the scale ; the obligation to conform to which, is of course common to every key. It would differ from the scale (supposing the mode to be major) in respect of the note F, the 4th of the original key of C, and the 7th of the new key. For, on recurring to the key of C as arranged according to the scale§, it will be found that, of the two semitones comprised in the scale, one falls between the 3rd and 4th notes, viz. between E and F, and the other between the 7th and 8th, viz. between B and C̅ ; but, in the new key of G, the interval between the 7th and

---

* See Appendix, Note (C).   † See Appendix, Note (D).

‡ It is obvious that, besides G̅,—C, D, E, and F also will each be an octave higher than the same note in the original key, and in strictness therefore should be marked C̅, D̅, E̅, F̅. But that particularity of notation would prove inconvenient for our present purpose.

§ Vide sup. p. 3.

8th, viz. between F and $\overline{\text{G}}$, would *not* be a semitone, but a whole tone; as will be apparent, by again recurring to the key of C†, where the interval from F to G is a tone. And the error here pointed out does not proceed from $\overline{\text{G}}$'s being too high; for $\overline{\text{G}}$, being a true octave to G, is right; but it proceeds from F's being too low; and it is F, therefore, that requires correction. To correct it, we must introduce, for the key of G, a different 7th, and one that shall be a semitone higher than F, so as to be a semitone only (as the scale requires) below $\overline{\text{G}}$. This new note might be denominated by a different letter; but it is more convenient to retain the F, and give it a mark expressive of its being a semitone higher than the F of the original key. It is, therefore, denominated *F sharp*, and marked *F*♯; for it is part of the method of notation in music, that this mark ♯ (called a *sharp*) raises the note, before or after which it is placed, a semitone; this mark ♭ (called a *flat*) lowers the note a semitone; and this mark ♮ (called a *natural*) restores it from flat or sharp to its original sound.

The new key will then be as follows:

G, A, B, C, D, E, F♯, $\overline{\text{G}}$;

or, according to the ordinary notation (in the treble clef),

the sharp being placed, not at the note F itself, but at the commencement of the staff, so as to point out the key more distinctly as the key of G, or one sharp,—for the key is denominated either way.

Upon the same principle, it will be found, that as we began with G, the 5th ascending from the original key note C, so, if the same progression of ascending 5ths be pursued, the

---

† Vide ibid.

effect will be to introduce at each new key one additional sharp, by way of correcting its 7th note. Thus :

Make D, the 5th from G, the key note, and its key will be D, E, F♯, G, A, B, C♯, D̄ ; or, according to the ordinary notation (in the treble clef),

and it is called the key of D, or two sharps. For as a sharp was added in the former case to F, the 4th note of the original key of C, in order to make it a correct 7th for the new key of G; so in like manner a sharp is now added to C, the 4th note of the key of G, in order to make it a correct 7th for the new key of D.

And so pursuing the same principle throughout, make A, the 5th from D, the key note, and its key will be A, B, C♯, D, E, F♯, G♯, Ā ; or, according to the ordinary notation (in the treble clef),

and it is called the key of A, or three sharps.

Make E, the 5th from A, the key note, and its key will be E, F♯, G♯, A, B, C♯, D♯, Ē ; or, according to the ordinary notation (in the treble clef),

and it is called the key of E, or four sharps.

Make B, the 5th from E, the key note, and its key will be B, C♯, D♯, E, F♯, G♯, A♯ B̄ ; or, according to the ordinary notation (in the treble clef),

and it is called the key of B, or five sharps.

Make the newly acquired F ♯ (which is the 5th from B) the key note, and its key will be F♯, G♯, A♯, B, C♯, D♯, E♯, F̄♯; or, according to the ordinary notation (in the treble clef),

and it is called the key of F sharp, or six sharps.

Make the newly acquired C♯ (which is the 5th from F♯) the key note, and its key will be C♯, D♯, E♯, F♯, G♯, A♯, B♯, C̄♯; or, according to the ordinary notation (in the treble clef),

and it is called the key of C♯, or seven sharps.

There being now seven sharps in the key (beyond which the notation by sharps cannot be carried, as a sharp is already applied to every note in the system), the progression by 5ths is abandoned; and the next resort is to the progression by 4ths. And, first, the note F (the 4th of the original key of C) is taken as the key note of a new key. If all the old notes were retained, thus—

F, G, A, B, C, D, E, F̄,

the note B, the 4th of the new key (being the 7th of the key of C) would then be erroneous; on the same principle as, in the key of G, the note F (the 7th of that key and 4th of the key of C) was erroneous. For, on recurring to the key of C, as arranged according to the scale*, we find that the 4th ought to be only a *semitone* above the 3rd; whereas B, as appears by the same arrangement, is a *whole tone* above A.

To correct this, therefore, the method is to *lower* B a semitone; introducing for that purpose a note called B flat (or B♭), upon the same principle that, in the key of G, the F was *heightened*, and became F sharp (or F♯). The new key will then be F, G, A, B♭, C, D, E, F̄; or, according to the ordinary notation (in the treble clef),

and it is called the key of F, or one flat.

And so, pursuing the same principle of progression by ascending 4ths, the effect will be to introduce, at each new key, one additional flat, by way of correcting, for the 4th of the new key, the 7th of the last preceding one. Thus:

Make B♭ (the 4th from F) the key note, and its key will be B♭, C, D, E♭, F, G, A, B̄♭; or, according to the ordinary notation (in the treble clef),

and it is called the key of B♭, or two flats.

Make E♭ (the 4th from B♭) the key note, and its key will be E♭, F, G, A♭, B♭, C, D, Ē♭; or, according to the ordinary notation (in the treble clef),

and it is called the key of E♭, or three flats.

---

* Vide sup. p. 3.

Make A♭ (the 4th from E♭) the key note, and its key will be A♭, B♭, C, D♭, E♭, F, G, A̅♭ ; or, according to the ordinary notation (in the treble clef),

and it is called the key of A♭, or four flats.

Make D♭ (the 4th from A♭) the key note, and its key will be D♭, E♭, F, G♭, A♭, B♭, C, D̅♭; or, according to the ordinary notation (in the treble clef),

and it is called the key of D♭, or five flats.

Make G♭ (the 4th from D♭) the key note, and its key will be G♭, A♭, B♭, C♭, D♭, E♭, F, G̅♭; or, according to the ordinary notation (in the treble clef),

and it is called the key of G♭, or six flats.

Make C♭ (the 4th from G♭) the key note, and its key will be C♭, D♭, E♭, F♭, G♭, A♭, B♭, C̅♭; or, according to the ordinary notation (in the treble clef),

and it is called the key of C♭, or seven flats; beyond which the notation by flats cannot be carried, as there is a flat already applied to every note in the system.

The result, therefore, on the whole, is that we have 15 keys in the major mode ; viz. the original key of C (which,

being without sharps or flats, is termed the *natural* key), 7
keys with sharps, and 7 with flats*. Thus :

$$
\begin{array}{ll}
\text{C} & \text{————— natural.} \\
\end{array}
$$

By ascending 5ths
$$
\left\{
\begin{array}{ll}
\text{G} & \text{————— 1 sharp.} \\
\text{D} & \text{————— 2 sharps.} \\
\text{A} & \text{————— 3 sharps.} \\
\text{E} & \text{————— 4 sharps.} \\
\text{B} & \text{————— 5 sharps.} \\
\text{F}\sharp & \text{————— 6 sharps.} \\
\text{C}\sharp & \text{————— 7 sharps.}
\end{array}
\right.
$$

By ascending 4ths.
$$
\left\{
\begin{array}{ll}
\text{F} & \text{————— 1 flat.} \\
\text{B}\flat & \text{————— 2 flats.} \\
\text{E}\flat & \text{————— 3 flats.} \\
\text{A}\flat & \text{————— 4 flats.} \\
\text{D}\flat & \text{————— 5 flats.} \\
\text{G}\flat & \text{————— 6 flats.} \\
\text{C}\flat & \text{————— 7 flats.}
\end{array}
\right.
$$

But of these, the keys with more than 5 sharps or flats are
seldom used.

In the *minor* mode†, the number of keys is the same, and
their arrangement is upon the same principle : though, in this
case, A (the 6th ascending from C) is taken (instead of C)
for the original note ; and the progression is first by ascending 4ths, and afterwards by ascending 5ths.

The key of A, when arranged according to the minor mode,
will be as follows (*descending*) :

A		G		F		E		D		C		B		A	
| tone |	tone	|semi|	tone	|	tone	|semi|	tone	|

or, acccording to the ordinary notation (in the treble clef),

and it is called the key of A *minor.*

It is to be observed, however, that, in the *ascending* scale
of any key in the minor mode, the highest interval is always
(as in the major mode) made a *semitone*, contrary to the proper

---

* See Appendix, Note (E).   † Vide sup. p. 2.

or *descending* progression of the scale, in which it is a *whole tone ;* and as this makes it necessary to augment the 7th note by a semitone, so a corresponding change is usually made on the 6th note also, in order to prevent the anomaly which would otherwise occur, of an interval *greater* than a tone between the 6th and 7th*.

Therefore, in A minor, *ascending*, the series will be

$$\text{A, B, C, D, E, F}\sharp\text{, G}\sharp\text{, } \overline{\text{A}} ;$$

or, according to the ordinary notation (in the treble clef),

the 6th, F $\sharp$, and the 7th, G $\sharp$, being *accidentals,* or foreign to the proper scale†.

In the particular key of A minor (that is, in its proper or *descending* series), it is also to be observed, that (as shewn in the last page) no $\sharp$ or $\flat$ requires to be introduced. It is consequently termed the *natural* key in that mode, as C is in the major. These keys of A minor and C major are also said to be *relative* to each other ; and so are all the keys in the major and minor modes respectively, which are thus connected with each other by similarity of notation—or (which is the same thing) comprise the same sounds differently arranged.

But in the other keys of the minor mode, flats or sharps require to be introduced, as in the case of the major mode ; the introduction, however, being at the 3rd or 7th note descending of each new key, instead of the 7th or 4th ascending. Thus, if we proceed by a progression first of ascending 4ths, and then of ascending 5ths (which in this mode is the most convenient order), the next key to A minor will have D for its key note, and will be as follows, differing (it will be observed) from A minor, by the introduction of a $\flat$ at B, the 3rd note descending :

(Descending)

D, C, B♭, A, G, F, E, D;

or, according to the ordinary notation (in the treble clef),

(Ascending)

D, E, F, G, A, B, C♯, D;

or, according to the ordinary notation (in the treble clef),

and it is called the key of D, or one flat minor.

And so of the rest.

The fifteen keys in the minor mode, according to the progression, first, of ascending 4ths, and next of ascending 5ths, will be found to be as follows :

A ———— natural (relative to C major).

By ascending 4ths.
{
D ———— 1 flat (relative to F major).
G ———— 2 flats (relative to B♭ major).
C ———— 3 flats (relative to E♭ major).
F ———— 4 flats (relative to A♭ major).
B♭ ——— 5 flats (relative to D♭ major).
E♭ ——— 6 flats (relative to G♭ major).
A♭ ——— 7 flats (relative to C♭ major).
}

By ascending 5ths.
{
E ———— 1 sharp (relative to G major).
B ———— 2 sharps (relative to D major).
F♯ ——— 3 sharps (relative to A major).
C♯ ——— 4 sharps (relative to E major).
G♯ ——— 5 sharps (relative to B major).
D♯ ——— 6 sharps (relative to F♯ major).
A♯ ——— 7 sharps (relative to C♯ major).
}

But, of these, E♭, A♭, D♯, A♯, are not generally used.

15

It may be useful here to subjoin a synopsis of the fifteen keys in each mode.

SYNOPSIS OF KEYS.

MAJOR MODE.

C	D	E	F	G	A	B	C
G	A	B	C	D	E	F♯	G
D	E	F♯	G	A	B	C♯	D
A	B	C♯	D	E	F♯	G♯	A
E	F♯	G♯	A	B	C♯	D♯	E
B	C♯	D♯	E	F♯	G♯	A♯	B
F♯	G♯	A♯	B	C♯	D♯	E♯	F♯
C♯	D♯	E♯	F♯	G♯	A♯	B♯	C♯
F	G	A	B♭	C	D	E	F
B♭	C	D	E♭	F	G	A	B♭
E♭	F	G	A♭	B♭	C	D	E♭
A♭	B♭	C	D♭	E♭	F	G	A♭
D♭	E♭	F	G♭	A♭	B♭	C	D♭
G♭	A♭	B♭	C♭	D♭	E♭	F	G♭
C♭	D♭	E♭	F♭	G♭	A♭	B♭	C♭

MINOR MODE.

Descending	A	G	F	E	D	C	B	A
Ascending	A	B	C	D	E	F♯	G♯	A
Descending	D	C	B♭	A	G	F	E	D
Ascending	D	E	F	G	A	B	C♯	D
Descending	G	F	E♭	D	C	B♭	A	G
Ascending	G	A	B♭	C	D	E	F♯	G
Descending	C	B♭	A♭	G	F	E♭	D	C
Ascending	C	D	E♭	F	G	A	B	C
Descending	F	E♭	D♭	C	B♭	A♭	G	F
Ascending	F	G	A♭	B♭	C	D	E	F
Descending	B♭	A♭	G♭	F	E♭	D♭	C	B♭
Ascending	B♭	C	D♭	E♭	F	G	A	B♭
Descending	E♭	D♭	C♭	B♭	A♭	G♭	F	E♭
Ascending	E♭	F	G♭	A♭	B♭	C	D	E♭
Descending	A♭	G♭	F♭	E♭	D♭	C♭	B♭	A♭
Ascending	A♭	B♭	C♭	D♭	E♭	F	G	A♭
Descending	E	D	C	B	A	G	F♯	E
Ascending	E	F♯	G	A	B	C♯	D♯	E
Descending	B	A	G	F♯	E	D	C♯	B
Ascending	B	C♯	D	E	F♯	G♯	A♯	B
Descending	F♯	E	D	C♯	B	A	G♯	F♯
Ascending	F♯	G♯	A	B	C♯	D♯	E♯	F♯
Descending	C♯	B	A	G♯	F♯	E	D♯	C♯
Ascending	C♯	D♯	E	F♯	G♯	A♯	B♯	C♯
Descending	G♯	F♯	E	D♯	C♯	B	A♯	G♯
Ascending	G♯	A♯	B	C♯	D♯	E♯	F♯♯	G♯
Descending	D♯	C♯	B	A♯	G♯	F♯	E♯	D♯
Ascending	D♯	E♯	F♯	G♯	A♯	B♯	C♯♯	D♯
Descending	A♯	G♯	F♯	E♯	D♯	C♯	B♯	A♯
Ascending	A♯	B♯	C♯	D♯	E♯	F♯♯	G♯♯	A♯

Such is the manner in which the several keys are constructed; a subject which we shall close by some remarks in reference to the *relation* which exists between keys.

A nearer relation obviously obtains between some keys than between others. The nearest is that which obtains between keys commonly described as "*relative*" to each other*; viz. a major and a minor key having altogether the same notes; and the next nearest is that between keys which differ only by a single ♯ or ♭.

And here the following rules will prove practically useful for the discovery of such keys as are thus related.

1. Any two keys in *different* modes will always be found to be *relative* to each other, if the key note of the minor is the 6th note of the major; as the keys of C major and A minor (where the key note of the latter is the 6th of the former).

2. Any two keys in *different* modes will always be found to differ only by a single ♯ or ♭, if the key note of the minor is the 2nd or 3rd of the major; as C major and D minor (where the key note of the latter is the 2nd of the former); or of F major and A minor (where the key note of the latter is the 3rd of the former).

3. Any two keys in the *same* mode will always be found to differ only by a single ♯ or ♭, if the key note of one is the 5th of the other; as the keys of C and G (where the key note of the latter is the 5th of the former).

We shall conclude the chapter by giving some account of that important part of the system of music which is commonly expressed by the word *Time,* and the closely connected subject of *Accent.*

Time regards either the *absolute* duration of sound allowed to different notes; or their *relative* duration as compared to one another; or the number of notes of the same relative duration respectively comprised in the different divisions of a piece of music.

On the *absolute* duration of the notes depends the character

---

* Vide sup. p. 13.

c

of a piece of music, as *quick* or *slow ;* and on this not much requires to be said, the time in this sense being according to the fancy of the composer, who points it out by such words, placed at the beginning of the movement, as are contained in the following table, expressing the different gradations from slow to quick.

Largo.

Larghetto.

Adagio.

Andantino.

Andante.

Allegretto.

Allegro.

Presto.

Prestissimo.

On the *relative* time of the notes it is also unnecessary to dwell, as it was suffieiently explained in a former place*, where the different forms and denominations of the notes, and their duration as compared with each other, were set forth. But time, in the third sense, requires more consideration, and it may be explained as follows :

Every piece of music is divided into small sections, called

bars, drawn across the stave, thus

each of which is to contain an equal number of notes of the same relative duration ; for example—four crotchets or three quavers; the time of the piece, if this number be divisible by two (as in the case of four crotchets) being called *common time ;* if divisible by three and not by two (as in the case of three quavers), *triple time.* And this contains the essence of the matter in a few words. But these other subordinate points are also to be taken into account.

First. Instead of any of the limited number of notes of given duration, any bar may comprise the equivalent of such notes. Thus, instead of four crotchets, a bar may comprise

---

two minims, which are equivalent; or three crotchets and two quavers, which are equivalent.

Secondly. If a dot be placed after any note, it is lengthened by half its value. Thus, a dotted crotchet ♩. is as long as a crotchet and a quaver, or three quavers.

Thirdly. Instead of any of the notes of given duration, its *rest* may be taken; that is, a pause, in which no note is sounded, but which lasts as long as a note of the given duration would have lasted, supposing it to have been sounded. Thus, instead of four crotchets, a bar may comprise three crotchets and a crotchet *rest*. And here it may be proper to set forth the different signs by which the rests are expressed, which are as follows:

Semibreve Rest...........

Minim Rest...............

Crotchet Rest. ....

Quaver Rest...............

Semiquaver Rest ....

Demisemiquaver Rest...

As to which, it is further to be understood, that the *rest* of a note, as well as the note itself, may be *dotted,* and with the same effect, viz. that it makes the *rest* one half longer.

Fourthly. If three notes have a curve drawn over them,

with the number 3 included, thus  which is called

a *triplet*, these notes are by a sort of license performed in the same time as two notes of the same value would be ; and if successive groups of three quavers occur, thus, each group is played in the time of one crotchet.

Common time, in its simple form, contains four crotchets, or their equivalent, and is marked thus at the beginning of the movement, ⌶ ₵ * ; or it contains two crotchets, or their equivalent (often called *half time*), and it is marked thus at the beginning of the movement: ²⁄₄

And here it is to be observed, that in all cases, where figures are thus used in reference to time, the lower one, if a 2, indicates *minims ;* if a 4, *crotchets ;* if an 8, *quavers ;* if a 16, *semiquavers* (these being expressive of the several proportions borne by these notes to a semibreve) ; and the upper figure shews how many of such notes are contained in a bar, or (as it is otherwise expressed) shews the number of *parts* that a bar contains.

Triple time, in its simple form, contains either three minims, three crotchets, or three quavers (or their equivalents) in a bar, and is marked ³⁄₂ ³⁄₄ ³⁄₈ in these several cases respectively.

Common time, however, is sometimes in a compound form, called *compound common time ;* and contains six crotchets in a

---

* There is also in occasional use the time called *Allabreve*, thus ₵ which is a quicker movement in common time. See Callcott's Grammar, p. 30.

bar, six quavers in a bar, or twelve quavers in a bar, and is marked $\frac{6}{4}$ $\frac{6}{8}$ $\frac{12}{8}$ in these cases respectively*.

Triple time is also sometimes in a compound form, which is called *Compound triple time ;* and contains nine crotchets, nine quavers, or nine semiquavers in a bar, and is marked $\frac{9}{4}$ $\frac{9}{8}$ $\frac{9}{16}$ in these cases respectively†.

Of the more usual of the several times above marked, the following are examples, and they exhibit also the manner of *counting* or *beating time*—methods by which the ear is guided in giving the proper duration to the notes, whether absolutely or relatively considered.

SIMPLE COMMON TIME.

Count or beat—

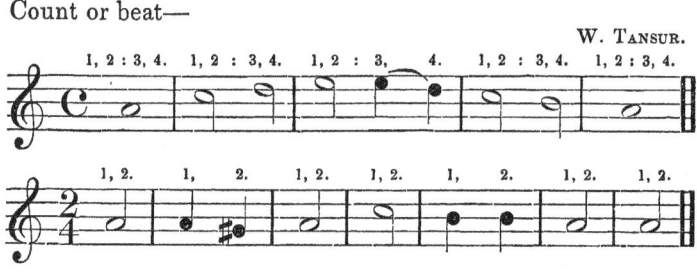

SIMPLE TRIPLE TIME.

* See Appendix, Note (H).   † See Appendix, Note (I).

COMPOUND COMMON TIME.

COMPOUND TRIPLE TIME.

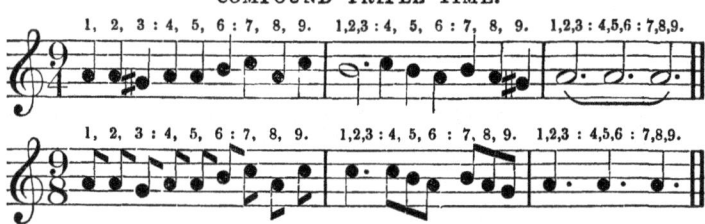

Accent consists in laying stress upon certain *parts* of a bar, as compared with the other *parts ;* and the observance of it is essential to the expression and general effect of musical performance.

In time marked $\mathbf{C}$, the accent is on the first and third parts of the bar, that is (supposing the time to be counted or beat), on numbers 1 and 3.

In $\frac{2}{4}$ time, the accent is on the first.

In $\frac{3}{2}$, $\frac{3}{4}$, $\frac{3}{8}$ time, the accent is on the first; or, according to some authorities, on the first and third, though less strongly on the third than on the first.

In $\frac{6}{4}$ or $\frac{6}{8}$ time, the accent is on the first and fourth parts.

In $\frac{9}{4}$ or $\frac{9}{8}$ time, the accent is on the first, fourth, and seventh.

The accented parts are usually called the *strong* parts in a bar; the unaccented, the *weak*.

CHAPTER II.

OF CHORDS IN GENERAL.

A chord is a combination of any number of notes struck together.

That department of music which relates to chords and their progression is called *Harmony ;* while the term *Melody* it used to express that which relates to a succession of single notes. Both terms, however, are also applied in a somewhat different way. For that succession of single notes which constitutes the leading part or *air,* is also technically called the *melody* of the piece ; and so, when the notes of a melody have chords put to them, the chords are said to be the *harmony* (or *accompaniment*) to the air or melody.

The notes of the scale, when the subject of harmony is concerned, usually receive denominations different from those in general belonging to them. The 1st or key note is in that case called the *tonic* (a term, however, precisely coincident in meaning with the *key note*) ; the 5th is called the *dominant ;* the 4th the *subdominant ;* and the 7th the *leading note ;* terms connected with their several properties or functions in the system of harmony*.

---

* See Appendix, Note (K).

In proceeding further to discuss the subject of chords (or *harmony*), we shall first consider chords *in general;* next, treat of their *progression ;* and, lastly, take a view of the *principal chords individually considered.* But, as introductory to the whole subject, we must in the first place enter into some further investigation of *Intervals ;* of which only those have been hitherto mentioned which lie between two *contiguous* notes of the scale*; the much larger number, contained between notes *not* contiguous, not having been yet noticed.

The following table exhibits examples of all those, of both kinds, which lie within the limits of a single octave (called *simple* intervals).

TABLE OF THE SIMPLE INTERVALS OF THE SCALE.

A *Semitone.* ....

A *Second* (or *tone*). ....

A *Minor Third*† (consisting of a tone and a semitone). ....

A *Major Third* (two tones). . ...

A *Fourth* (two tones and a semitone). ....

A *Tritone* (three tones). ....

A *Minor Fifth* (two tones and two semitones). ..

---

* Vide sup. p. 2.   † See Appendix, Note (L).

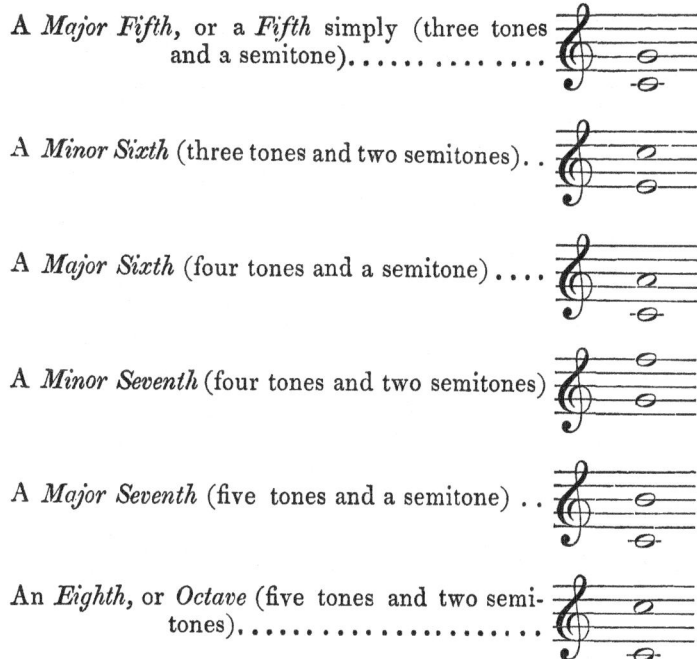

A *Major Fifth*, or a *Fifth* simply (three tones and a semitone)..............

A *Minor Sixth* (three tones and two semitones)..

A *Major Sixth* (four tones and a semitone)....

A *Minor Seventh* (four tones and two semitones)

A *Major Seventh* (five tones and a semitone)..

An *Eighth*, or *Octave* (five tones and two semitones).....................

A second class of intervals consist of the former altered, by having the higher of their constituent notes raised, or the lower depressed, one octave, or several octaves; so that an interval of this class may be said to be *compounded* of an octave, double octave, &c. with one of the other intervals of the former class; and the distinction between the two is accordingly marked, by giving the name of *simple* intervals to the former class, and *compound* intervals to the latter. Examples of compound intervals are exhibited in the following Table.

TABLE OF COMPOUND INTERVALS OF THE SCALE.

A *Ninth* (minor), being an octave compounded with a semitone.............

A *Ninth* (major), being an octave compounded with a second..............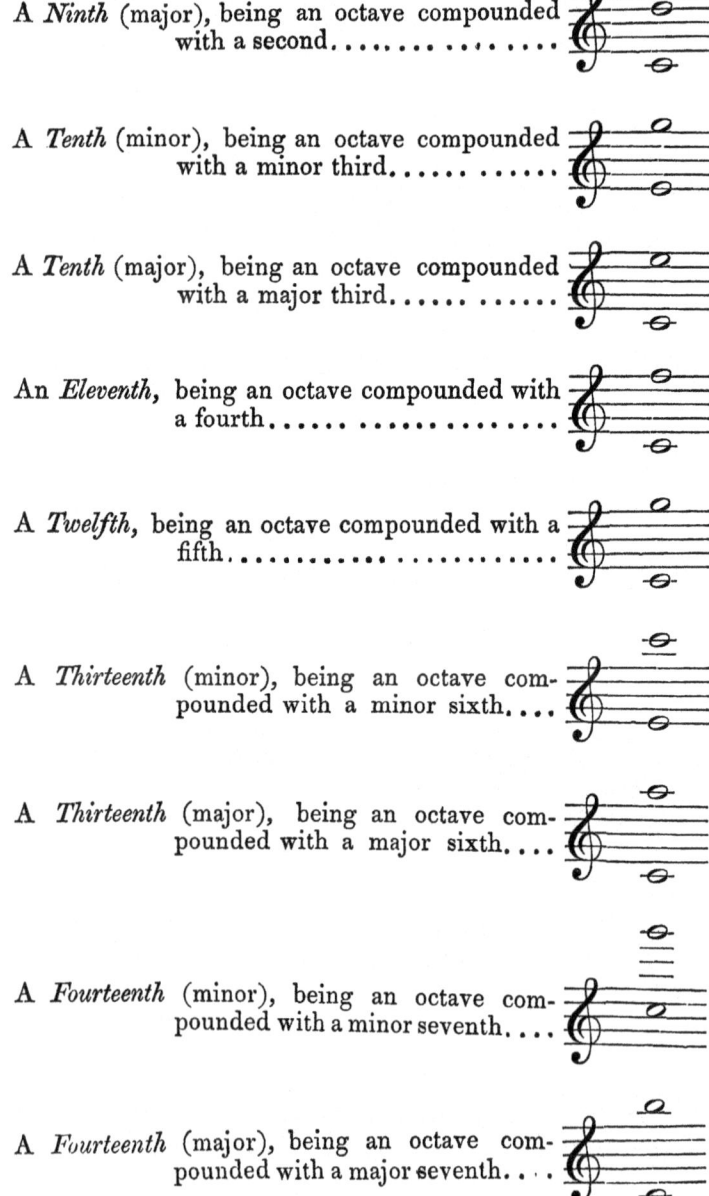

A *Tenth* (minor), being an octave compounded with a minor third...... ......

A *Tenth* (major), being an octave compounded with a major third...... ......

An *Eleventh*, being an octave compounded with a fourth..... ....

A *Twelfth*, being an octave compounded with a fifth.................. ....

A *Thirteenth* (minor), being an octave compounded with a minor sixth....

A *Thirteenth* (major), being an octave compounded with a major sixth....

A *Fourteenth* (minor), being an octave compounded with a minor seventh....

A *Fourteenth* (major), being an octave compounded with a major seventh....

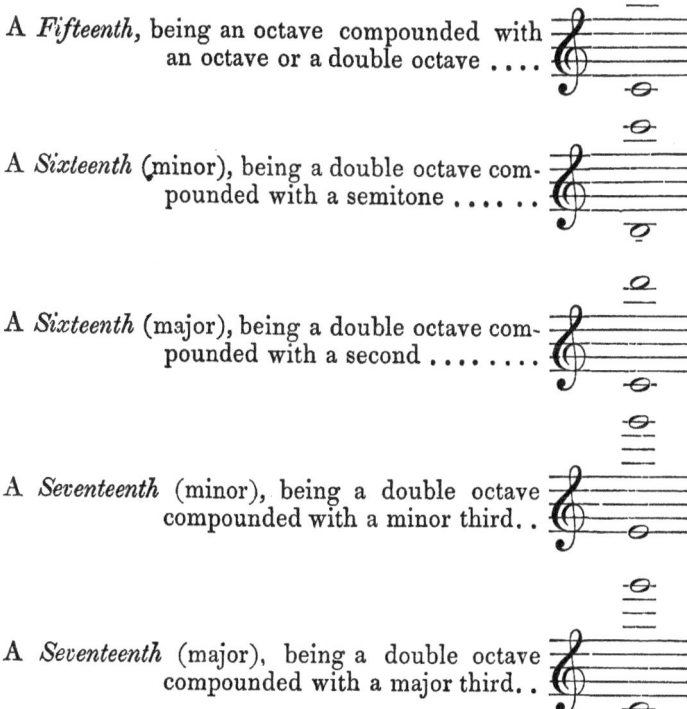

A *Fifteenth*, being an octave compounded with an octave or a double octave ....

A *Sixteenth* (minor), being a double octave compounded with a semitone ......

A *Sixteenth* (major), being a double octave compounded with a second ....

A *Seventeenth* (minor), being a double octave compounded with a minor third..

A *Seventeenth* (major), being a double octave compounded with a major third..

Besides these, there is a third class of intervals *foreign* to the scale, and introduced by way of variation upon it ; and to which we may assign the denomination of *Extraneous* intervals.

These are—

1. *The Enharmonic interval**.

This is the least interval recognised in music, and is that which exists between two contiguous notes, one marked ♯, and the other ♭ ; for example, between D♯ and E♭ . But in keyed instruments, that is, instruments with fixed keys or frets,

---

* Sometimes called (though incorrectly) a *Quartertone*.

and which are consequently tuned by *temperament* (as to which, vide post. Chap. V), this interval has in fact no existence, the same sound serving to express the sharp of the lower and the flat of the upper note ; though the distinction between these is nevertheless of importance, and is still preserved in point of notation.

2. *The Chromatic (or minor) semitone.*

This is the interval which exists between any note in the scale and the same note raised by a ♯ or lowered by a ♭ ; for example, the interval or difference between C and C♯, or between E and E flat ; differing in this respect from the semitone *of the scale* (called, by way of distinction from it, the *diatonic* semitone)*, which is an interval between different notes of the scale, viz. between the 3rd and 4th, and between the 7th and 8th ; for example, between E and F, or between B and C, in the key of C major† ; and which is also a somewhat larger interval than the *chromatic* semitone ; though, in instruments with fixed keys or frets, these intervals are equalized by *temperament,* in the tuning.

3. *Certain of the proper intervals of the scale, either augmented or diminished by a chromatic semitone.*

As when the tone C D is augmented into C D♯ by the addition of the chromatic semitone from D to D♯ ; or when the minor 7th, G F, is diminished into G♯ F by the subtraction of the *chromatic* semitone from G to G♯‡.

Of these three kinds of extraneous intervals, examples are set forth in the following Table, which comprises all the principal intervals of that class§.

TABLE OF EXTRANEOUS INTERVALS.

An *Enharmonic interval*....................

---

* See Appendix, Note (M).   † Vide sup. p. 3.
‡ See Appendix, Note (N).   § See Appendix, Note (O).

A *chromatic Semitone*. . . . .

An *augmented 2nd*. . . . .

A *diminished minor 3rd*. . . . .

An *augmented major 3rd*. . . . .

A *diminished 4th*. . . . .

An *augmented 4th*. . . . .

A *diminished 5th*. . . . .

An *augmented 5th*. . . . .

A *diminished minor 6th*. . . . .

An *augmented major 6th (or extreme sharp 6th)*

A *diminished minor 7th.* . . . .

Though a chord has been correctly above defined as a combination of any number of notes struck together, yet, in its most proper sense, it consists of four notes (or three

intervals), thus

It often consists, nevertheless, of fewer notes, viz. of three,

thus 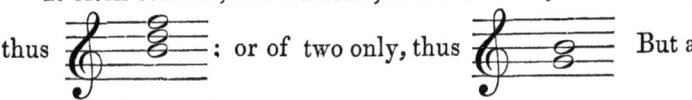 : or of two only, thus But a

chord of only two notes (or one interval) is usually described, not as a chord, but simply as a 3rd, 5th, &c. according to the nature of the interval.

A chord may, on the other hand, consist of *more* than four notes. But in such case the additional sounds are usually mere octaves or repetitions of some of the others ; so that the sounds are in effect but four. There are instances, however, of chords which contain more than four distinct sounds.

The intervals of a chord are all counted upwards, from the *bass* (or *lowest* note), and it takes its denomination from the nature of its intervals so counted. Thus, if a chord contain the notes G, B, D, F, its intervals counted from the bass note are G B, G D, G F, which are a major 3rd, a 5th, and a minor 7th ; and it is consequently called a chord of the seventh, fifth, and third, or (more shortly) a chord of the seventh ; and, where the chords are distinguished by figures (as is often

the case), is figured 5, or 7.
$$\begin{array}{c} 7 \\ 5 \\ 3 \end{array}$$

If the order of the notes of any chord be so changed as to involve a change of the bass note, its intervals and denomination will be changed*. Thus, if the chord G, B, D, F, be changed in the order of its notes, by making B the bass

---

* See Appendix, Note (P).

note, and turned into B, G, D, F, its intervals will now be B G, B D, B F, which are a minor 6th, a minor 3rd, and a minor 5th, instead of a major 3rd, 5th, and minor 7th; and its denomination will be a chord of the sixth, fifth, and third, or a chord of $\frac{6}{5}$, or $\frac{6}{5}$. On the other hand, if the order of the notes be changed *without* changing the bass note, the intervals and denomination will remain the same as before. Thus, if the chord G, B, D, F, be changed into G, D, B, F, the intervals will be G D, G B, G F, which are the same as those of the original chord, G, B, D, F; and its denomination will still be a chord of the seventh, fifth, and third, or of the seventh; or a chord of $\frac{7}{5}$, or 7. In the first case, viz. that of a change in the bass note, the new chord thereby produced is considered as a *different* chord from the original one, because the intervals are altered; but in the second, where the new arrangement involves no change in the bass note, or in the intervals, the chord produced is considered as substantially the *same* with the original one. In either case, however, it will be observed, that there is so far an identity between the new chord and the original one, that both consist of exactly the same notes.

There is one chord which far exceeds all others in importance, and is indeed the basis of the whole system of Harmony. This is called the *common chord major*, or simply the *common chord**. It consists of the intervals of a major 3rd, a 5th, and an 8th; thus ; and it is figured $\frac{8}{5}$, or $\frac{5}{3}$, or (more usually) *no* figuring is used; the rule being—that when, among chords that are figured, any occur

---

* See Appendix, Note (Q).

to which no figure is applied, a chord with the intervals of 3rd, 5th, and 8th, is intended.

When the order of the notes of a common chord major is so changed as to involve a change of the bass note, this chord is said to be *inverted ;* and the new chord thereby produced is called an *inversion* of the common chord major; and this last, as distinguished from its inversions, is called the common chord major *direct.* Thus, in the above example, the chord C, E, G, C̄, is a common chord major direct; of which E, C, G, C̄ would be an inversion. But when the order of the notes of the common chord major is changed, *without* changing the bass note, the chord thereby produced is called, not an inversion, but a new *position* of the common chord major. Thus, C, G, E, C̄ would be a new position, not an inversion, of the common chord major, C, E, G, C. It follows from this, in connection with what was before laid down*, that a common chord major inverted is a different chord from a common chord major direct; but that a common chord major, under any change of position, still remains substantially the same chord.

An important and curious fact relating to the common chord major, is that it is the *product of nature herself.* For it is found, by experiment, that any string or other body, capable of emitting a musical sound, yields, upon being struck, not only its principal note, but (in quick succession) these others also, viz. the 8th ascending, the 12th ascending, and the 17th major ascending, of its principal note†; in other words, it yields notes comprising the common chord major of the principal note; for if the 12th and 17th be lowered, the first an octave and the second a double octave, they become the 5th and 3rd respectively of the principal note ; and the sounds then consist of the principal note, with its 3rd, 5th, and 8th, which, as we have seen, constitute the common chord major.

---

* Vide sup. p. 31.   † See Appendix, Note (R).

The bass note of a common chord major is, in reference to this fact, sometimes called, in theoretical disquisition, the *generator* of the three others, viz. the 3rd major, the 5th, and the 8th ; and these others are frequently termed its *harmonics ;* but the bass is otherwise (and more frequently) termed the *fundamental bass* (or *root*) of the several other notes, and it is also described as the fundamental bass or root of the whole chord taken together, and of all its inversions. Thus C, the bass note of the common chord major, C, E, G, $\overline{\text{C}}$, is called the fundamental bass or root of each of the notes, E, G, $\overline{\text{C}}$, and of the whole chord C, E, G, $\overline{\text{C}}$, and also of its inversions, viz. E, G, C, $\overline{\overline{\text{E}}}$, and G, $\overline{\text{C}}$, E, G.

Besides the common chord major, there are three others to which particular attention is due. Two of them are generically connected with the common chord major, because consisting, like it, of 3rd, 5th, and 8th ; viz. the *common chord minor,* for example, , consisting of a *minor* 3rd, 5th, and 8th ; and the *imperfect common chord,* for example, , consisting of a *minor* 3rd, a minor 5th, and an 8th*. The remaining chord is the *chord of the 7th,* for example, , consisting of 3rd, 5th, and 7th ; all which intervals may be either major or minor. The first two of these chords follow the same rule, as to figuring, as the common chord major ; the last is figured $\begin{smallmatrix}7\\5\\3\end{smallmatrix}$, or, more simply, 7.

The same distinction, as to *inversions* and *positions*, applies

---

* See Appendix, Note (S).

to these chords, as to the common chord major. For any of these, as well as the common chord major, is said to be *inverted*, when its bass note is changed; and, in any other case, to be *direct**. It is also said, in the case of a new arrangement not involving a change of the bass note, to be in a new *position;* and, when in a new position, it remains substantially the same chord as before; but, when inverted, becomes a different chord.

As regards these chords also, as well as the common chord major, it is found convenient, for the purposes of arrangement, to consider the bass note as constituting the *fundamental bass* of the chord (whether direct or inverted), and of its several notes. It is true that the bass notes of these chords do not strictly satisfy the term of fundamental bass, according to its proper meaning; for they can never *generate* the whole chords of which they are the basses—no note being capable of generating, as its harmonics, its minor 3rd, or minor 5th, or 7th, or any notes, except its major 3rd, 5th, and 8th. But, in a secondary and less proper sense, the bass notes of such chords may be considered as fundamental basses; viz. as standing, in relation to the whole chord, as 1 to 3, 5, 8, or 1 to 3, 5, 7. In this sense, we may consider D (the bass note of the common chord minor, D, F, A, D) and C (the bass note of the chord of the 7th, C, E, G, B) as the fundamental basses also of those chords respectively (whether direct or inverted), and of their constituent notes†.

In order to obtain a just and comprehensive view of the principles of harmony, it is of the first importance to understand the constitution of the scale, in reference to the several chords which have been above particularized.

Confining our attention then, in the first instance, to the common chords (major, minor, and imperfect), and considering them only as they occur in the major mode, we shall find that in every key there are seven chords of this nature, of which the several notes of the key constitute respectively the bass

---

* See Appendix, Note (T).   † See Appendix, Note (U).

notes, and therefore (as results from explanations already given) the fundamental basses.

Thus, of the seven notes of which the key consists (for the 8th being a mere replicate of the tonic, and having consequently the same incidents, requires no separate consideration), the *tonic, subdominant,* and *dominant* may be made to carry a common chord *major ;* the *2nd, 3rd,* and *6th,* a common chord *minor ;* and the *leading note,* an *imperfect* common chord. For example, in the key of C major, the tonic C may be made to carry the common chord, C, E, G, $\overline{\text{C}}$; the subdominant F, the common chord, F, A, $\overline{\text{C}}$, $\overline{\text{F}}$; the dominant G, the common chord, G, B, $\overline{\text{D}}$, $\overline{\text{G}}$; the 2nd D, the common minor, D, F, A, $\overline{\text{D}}$; the 3rd E, the common chord minor, E, G, B, $\overline{\text{E}}$; the 6th A, the common chord minor, A, $\overline{\text{C}}$, $\overline{\text{E}}$, $\overline{\text{A}}$; and the leading note B, the imperfect common chord, B, $\overline{\text{D}}$, $\overline{\text{F}}$, $\overline{\text{B}}$; all which it may be convenient to exhibit in a tabular form.

THE COMMON CHORDS OF THE KEY.

(In the major mode. In the key of C.)

	$\overline{\text{C}}$	$\overline{\text{D}}$	$\overline{\text{E}}$	$\overline{\text{F}}$	$\overline{\text{G}}$	$\overline{\text{A}}$	$\overline{\text{B}}$
	G	A	B	$\overline{\text{C}}$	$\overline{\text{D}}$	$\overline{\text{E}}$	$\overline{\text{F}}$
	E	F	G	A	B	$\overline{\text{C}}$	$\overline{\text{D}}$
Bass Notes.	C	D	E	F	G	A	B
	Com. Chord, Major.	Com. Chord, Minor.	Com. Chord, Minor.	Com. Chord, Major.	Com. Chord, Major.	Com. Chord, Minor.	Com. Chord, Imper.

Here, each column contains a common chord major, minor, or imperfect, of which the several notes of the key are respectively the bass notes and fundamental basses*.   And the key affords no other but these ; for we cannot take a common chord, other than those enumerated, without going out of the key. Thus, in the key of C major, if we take D for a bass note, on which to construct a common chord *major*, instead of a common chord *minor*, there will be no 3rd major for the purpose, without going out of the key and taking F$\sharp$; for F$\natural$, which is the only F belonging to the key, makes with D a 3rd minor only.   And a difficulty of the same kind will be found, if we attempt to make the 3rd, the 6th, or the leading note carry a common chord major ; or the tonic, sub-dominant, or dominant carry any common chord but a major one.

Those notes of any key which constitute the fundamental basses of common chords *major*, viz. the *tonic, subdominant,* and *dominant,* we shall denominate the *primary ;* and those which constitute the fundamental basses of common chords *minor* and *imperfect,* viz. the *2nd, 3rd, 6th,* and *leading note,* the *secondary* fundamental basses of the key ; it being by the harmonies of the former that the key is chiefly characterized, and most distinctly impressed upon the ear†.

It will consequently be apparent that every note, in any given key, has its proper fundamental basses, both primary and secondary ; and that some of the notes have more than one primary, and others more than one secondary fundamental bass.   Thus, $\overline{C}$ in the first column of the table, p. 35, has C in the same column, or F in the fourth column, for its primary, and A in the sixth column for its secondary, fundamental bass. And so of all the remaining notes in the key ; as will be more fully illustrated by the following Table, taken in connection with the former one.

---

* See Appendix, Note (V).
See Appendix, Note (W).

Notes of the Key	C̄	D̄	Ē	F̄	Ḡ	Ā	B̄
Primary....	C	G	C	F	C	F	G
Ditto	F				G		
Secondary..	A	D	E	D	E	A	B
Ditto		B	A	B		D	E

Fund. Basses.

While the major mode admits (as we have seen) a common chord major on the tonic, subdominant, and dominant, a common chord minor on the 2nd, 3rd, and 6th notes, and an imperfect common chord on the leading note, the case is different with the *minor* mode, which, according to its scale in the proper or *descending* progression, admits a common chord minor on the tonic, subdominant, and dominant, a common chord major on the 2nd, 3rd, and 6th descending, and an imperfect common chord on the 7th descending. But, according to its scale in the *ascending* progression (which augments two of the notes by a chromatic semitone*), a common chord major may also be taken on its subdominant and dominant; and, in practice, the common chord of the *dominant* in this mode (as well as in the other) is *invariably* taken major†.

Having thus described the constitution of the scale in reference to the common chords major, minor, and imperfect, its constitution, in reference to the chord of the 7th, may be very briefly noticed, being in fact precisely similar, as may be sufficiently inferred from the consideration that a chord of the 7th is no other than a common chord major, minor, or imperfect, with a 7th (major or minor) added. But it may be acceptable to illustrate this subject, as we did the former, in a tabular point of view.

---

* Vide sup. p. 13.

† See Appendix, Note (X).

THE SEVENTHS OF THE KEY.
(In the major mode. In the key of C.)

B	C̄	D̄	Ē	F̄	Ḡ	Ā
G	A	B	C̄	D̄	Ē	F̄
E	F	G	A	B	C̄	D̄
C	D	E	F	G	A	B
Chord of the 7th with Major 3rd, 5th and Major 7th.	Chord of the 7th with Minor 3rd, 5th and Minor 7th.	Chord of the 7th with Minor 3rd, 5th and Minor 7th.	Chord of the 7th with Major 3rd, 5th and Major 7th.	Chord of the 7th with Major 3rd, 5th and Minor 7th.	Chord of the 7th with Minor 3rd, 5th and Minor 7th.	Chord of the 7th with Minor 3rd, Minor 5th, and Minor 7th.

After these explanations in regard to certain chords of primary importance, and the constitution of the scale in regard to them, it is now time to apply ourselves to the subject of chords in general.

All chords range themselves under the different classes of *Concords and Discords ;* the distinction between which will be best understood, by first adverting to the case of chords of only two notes ; or, in other words, the notes of any *interval* sounded together.

Chords of two notes, then, are said to be *concords,* when they are comprised within some common chord major or minor ; and are said to be *discords,* when they are not comprised within any common chord major or minor.

According to this definition, seven of the simple intervals of the scale*, viz. the 3rd (major and minor), the 4th, the 5th, the 6th (major and minor), and the 8th, are concords† ; for each of them occurs in a common chord, major or minor. Thus the major 3rd, C E, the minor 3rd, E G, the 4th, G C̄, the 5th, C G, the minor 6th, E C̄, and the 8th, C C̄, all occur in the com-

---

* Vide sup. p. 24.　　　　† See Appendix, Note (Y).

mon chord major, C, E, G, $\overline{\text{C}}$; and the major 6th, $\overline{\text{G} \ \text{E}}$, occurs in the same chord taken in the position, C, G, $\overline{\text{C}, \text{E}}$; but the remaining six of the simple intervals of the scale, viz. the semitone, the 2nd, the tritone, the minor 5th, and the 7th major and minor are discords; for none of these occur in any common chord major or minor. According to the same definition, too, such of the compound intervals of the scale*, as consist of an octave compounded with concords, are themselves also concords; and such of them as consist of an octave compounded with discords, are discords. Thus the 10th minor, $\overline{\text{E} \ \text{G}}$, which is compounded with the concord of the minor 3rd, E G, is a concord; but the 9th major, $\overline{\text{C} \ \text{D}}$, which is compounded with the 2nd, C D, a discord. And, lastly, according to the same definition, all extraneous intervals† are discords; for, being foreign to the scale, they are of course foreign also to the common chord, which belongs to the scale.

Not only the combinations of *two* notes are said to be discords or concords, but the terms are applied to chords in the fuller sense: viz. those comprising *more* than two notes; such of them as comprise not any discordant interval being concords, and such as do, being discords. Thus the entire common chord major, C, E, G, C (as well as the concordant intervals of the 3rd, 5th, and octave of which it consists), is called a concord; and the chord of the 7th, G, B, D, F (as well as the discordant intervals of the minor 7th, G F, and the minor 5th, B F, which it comprises), is called a discord; being made so by these, its discordant parts; though it also comprises concordant intervals, viz. the major 3rd, G B, the fifth, G D, and the minor 3rds, B D and D F. It is to chords consisting of more than two notes, that the terms of *Concords* and *Discords* are in future to be understood to apply, unless the contrary shall be expressed.

The concords all afford unqualified pleasure to the ear. The discords, on the other hand, are all attended with a sort of jarring or confusion, of which the effect is (in itself) un-

---

* Vide sup, p. 25. † Vide sup. p. 27.

pleasing, and by which they are generically distinguished from the concords. The discords, however, differ from each other very widely in their character. The effect of some (in which the jarring is less perceptible) is, on the whole, intrinsically pleasing; that of others (in which it is more perceptible), displeasing, except in their connection with other sounds; and that of others (in which it prevails still more decidedly), too harsh to be tolerated under any circumstances.

Of these kinds, the two first are freely admitted into musical composition, and perform services of the utmost importance; for, by way either of *variety* or *contrast*, they afford a grateful and necessary relief to the concords, an unbroken succession of which would soon prove cloying and wearisome to the ear*.

The concords are easily enumerated, for they are no other than the common chords major and minor, and their respective inversions; every common chord having two inversions; one called the *chord of the* 6*th* and 3*rd*, or (more shortly) *of the* 6*th*, comprising the intervals of 3rd, 6th, and 8th, and figured 6, or $\frac{8}{6}$; and the other called the *chord of the* 6*th* and 4*th*, comprising the intervals of 4th, 6th, and 8th, and figured $\frac{8}{6}$, or $\frac{6}{4}$. Examples, both of the two concords and their inversions, will be found in the subjoined Table.

The discords present, on the other hand, a wide and complicated field, such as cannot conveniently be traversed without the aid of some analysis. The following is that which has occurred to the writer of this work, and which has, at least, this recommendation—that it is founded not on any supposed harmonical relations between one chord and another in point of effect or use (always a very fallacious basis), but simply on the intervals which they respectively contain; a point on which there can be no mistake.

---

* See Appendix, Note (Z).

Discords are divisible into, I, such as consist of combinations that may be found without going out of the scale, and may therefore be termed DISCORDS OF THE SCALE; II, such as consist of combinations that cannot be found without going out of the scale, and may therefore be termed EXTRANEOUS DISCORDS.

I. As to the DISCORDS OF THE SCALE, they always comprise some discordant interval or intervals of the scale*; and such interval or intervals may either be formed between two upper notes of the chord, or between the bass and an upper note; or in both these ways. Thus the chord of the $\begin{smallmatrix}8\\5\\4\end{smallmatrix}$

comprises the discordant interval of a 7th, G F, formed between two upper notes; the chord of the $\begin{smallmatrix}6\\4\\2\end{smallmatrix}$ comprises the discordant interval of a 2nd, C D, formed between the bass and an upper note; and the chord of the $\begin{smallmatrix}7\\5\\3\end{smallmatrix}$ comprises both the discordant interval of a 7th, G F, formed between the bass and an upper note; and that of a minor 5th, B F, formed between two upper notes.

The plan that will be followed in this work for the arrangement of the Discords of the Scale, will be first to enumerate those which comprise no discordant interval but such as is formed between upper notes; and next, those which comprise discordant intervals formed between the bass and an upper note, whether also comprising such as are formed between upper notes or not; and to arrange the latter discords (which are much the most numerous) according to the order of the discordant intervals which they respectively form with the bass; viz.

---

* Vide sup. p. 39.

2nd, tritone, 5th, 7th, or 9th.   And the following is a Scheme of the principal Discords of the Scale, arranged according to this method.

1.   *Discords with the discordant interval of a 2nd (or 7th) formed between upper notes.*

$$\begin{array}{r}8\\5\\4\end{array}$$

8th, 5th, and 4th..

No. 7 in the Table of Chords, post, p. 53.

$$\begin{array}{r}6\\4\\3\end{array}$$

Major 6th, 4th, and minor 3rd............................

No. 13 in the Table.

$$\begin{array}{r}6\\5\\3\end{array}$$

Major 6th, 5th, and major 3rd............................

No. 16 in the Table.

$$\begin{array}{r}6\\4\\3\end{array}$$

Minor 6th, 4th, and minor 3rd............................

No 17 in the Table.

2.   *Discords with the discordant interval of a 2nd formed with the bass.*

$$\begin{array}{r}6\\4\\2\end{array}$$

Major 6th, 4th, and 2nd...................................

No. 18 in the Table.

$$\begin{array}{r}5\\2\end{array}$$

5th and 2nd...

No. 8 in the Table.

$$\begin{array}{r}5\\4\\2\end{array}$$

5th, 4th, and 2nd...

No. 9 in the Table.

3.   *Discord with the discordant intervals of a 2nd and a tritone formed with the bass.*

$$\begin{array}{r}6\\4\\2\end{array}$$

Major 6th, tritone, and 2nd...............................

No. 14 in the Table.

4. *Discords with the discordant interval of a minor 5th formed with the bass.*

8th, minor 5th, and minor 3rd (called *imperfect common chord*) . . . .
$$\begin{matrix} 8 \\ 5 \\ 3 \end{matrix}$$

No. 10 in the Table.

Minor 6th, minor 5th, and minor 3rd. . . . .
$$\begin{matrix} 6 \\ 5 \\ 3 \end{matrix}$$

No. 12 in the Table.

5. *Discords with the discordant interval of a 7th formed with the bass.*

Minor 7th, 5th, and major 3rd (called *chord of the dominant 7th*). . . . .
$$\begin{matrix} 7 \\ 5 \\ 3 \end{matrix}$$

No. 11 in the Table.

Minor 7th, 5th, and minor 3rd (called *chord of the 7th upon the 2nd of the key*). . . . .
$$\begin{matrix} 7 \\ 5 \\ 3 \end{matrix}$$

No. 15 in the Table.

Minor 7th, 4th, and minor 3rd. . . . .
$$\begin{matrix} 7 \\ 4 \\ 3 \end{matrix}$$

No. 19 in the Table.

Minor 7th, major 6th, and major 3rd. . . . .
$$\begin{matrix} 7 \\ 6 \\ 3 \end{matrix}$$

No. 20 in the Table.

Minor 7th, 5th, and 4th. . . . .
$$\begin{matrix} 7 \\ 5 \\ 4 \end{matrix}$$

No. 21 in the Table.

6. *Discords with the discordant intervals of a 7th and a 2nd formed with the bass.*

Major 7th, 4th, and 2nd. . . . .
$$\begin{matrix} 7 \\ 4 \\ 2 \end{matrix}$$

No. 22 in the Table.

Major 7th, major 6th, 4th, and 2nd............... ....
$$\begin{matrix} 7 \\ 6 \\ 4 \\ 2 \end{matrix}$$

No. 23 in the Table.

Major 7th, 5th, 4th, and 2nd......................
$$\begin{matrix} 7 \\ 5 \\ 4 \\ 2 \end{matrix}$$

No. 24 in the Table.

7. *Discords with the discordant interval of a 9th formed with the bass.*

Major 9th, 5th, and major 3rd (called *chord of the 9th*)..
$$\begin{matrix} 9 \\ 5 \\ 3 \end{matrix}$$

No. 25 in the Table.

Minor 9th, minor 6th, and minor 3rd...... ....
$$\begin{matrix} 9 \\ 6 \\ 3 \end{matrix}$$

No. 26 in the Table.

Minor 9th, minor 6th, and 4th,................ ......
$$\begin{matrix} 9 \\ 6 \\ 4 \end{matrix}$$

No. 27 in the Table.

Major 9th, 5th, and 4th. ....
$$\begin{matrix} 9 \\ 5 \\ 4 \end{matrix}$$

No. 28 in the Table.

8. *Discords with the discordant intervals of a 9th and a 7th formed with the bass.*

Major 9th, minor 7th, and minor 3rd.............. ....
$$\begin{matrix} 9 \\ 7 \\ 3 \end{matrix}$$

No. 29 in the Table.

Major 9th, major 7th, and 4th.................... $\begin{matrix} 9 \\ 7 \\ 4 \end{matrix}$

No. 30 in the Table.

II. The EXTRANEOUS DISCORDS (or those foreign to the scale) always comprise some one or more of those extraneous intervals which consist of proper intervals of the scale augmented or diminished by a chromatic semitone*; and such extraneous intervals may either occur between two upper notes of the chord, or between the bass and an upper note.

The following is a Scheme of some of the Extraneous Discords, comprising all such as are of primary importance.

1. *Discords with the extraneous interval of an augmented 2nd formed with the bass.*

Major 6th, tritone, and augmented 2nd............... $\begin{matrix} 6 \\ 4 \\ 2 \end{matrix}$

No. 43 in the Table.

Major 7th and augmented 2nd....................... $\begin{matrix} 7 \\ 2 \end{matrix}$

No. 31 in the Table.

2. *Discord with the extraneous interval of a diminished 4th formed with the bass.*

Minor 6th and diminished 4th...................... $\begin{matrix} 6 \\ \flat 4 \end{matrix}$

No. 32 in the Table.

3. *Discords with the extraneous interval of an augmented 4th formed with the bass.*

Major 6th, augmented 4th, and minor 3rd........... $\begin{matrix} 6 \\ 4 \\ 3 \end{matrix}$

No. 42 in the Table.

---

* Vide sup. p. 28, and Appendix, Note (AA).

4. *Discord with the extraneous interval of an augmented 5th formed between two upper notes.*

Minor 6th and major 3rd. . . . .　6
♯

No. 33 in the Table.

5. *Discord with the extraneous interval of an augmented 5th formed with the bass.*

8th, augmented 5th, and major 3rd. . . . .　5̶

No. 34 in the Table.

6. *Discord with the extraneous interval of an augmented major 6th formed between two upper notes.*

Minor 7th, minor 5th, and major 3rd . . . .　7
5
♯

No. 35 in the Table.

7. *Discords with the extraneous interval of an augmented major 6th formed with the bass.*

Augmented major 6th and major 3rd (called *chord of the extreme sharp 6th*). . . . .　6̶

No. 36 in the Table.

Augmented major 6th, 5th, and major 3rd. . . . .　6̶
5
3

No. 37 in the Table.

Augmented major 6th, tritone, and major 3rd. . . . .　6̶
4
3

No. 38 in the Table.

8. *Discord with the extraneous interval of a diminished minor 7th formed between two upper notes.*

Major 7th, minor 6th, 4th, and 2nd . . . .　7
6
4
2

No. 39 in the Table.

9. *Discord with the extraneous intervals of a diminished minor 7th and a diminished 5th formed between two upper notes.*

Major 6th, minor 5th, and minor 3rd................. $\begin{matrix} 6 \\ 5 \\ 3 \end{matrix}$

No. 41 in the Table.

10. *Discord with the extraneous intervals of a diminished minor 7th and a diminished 5th formed with the bass.*

Diminished minor 7th, diminished 5th, and minor 3rd, (called *chord of the diminished 7th*).......... $\begin{matrix} 7 \\ \text{or} \\ \flat 7 \end{matrix}$

No. 40 in the Table.

11. *Discord with the extraneous intervals of a diminished minor 7th, a diminished 5th, and a diminished minor 3rd, formed with the bass.*

Diminished minor 7th, diminished 5th, and diminished minor 3rd........................... $\begin{matrix} 7 \\ 5 \\ \flat \end{matrix}$

No. 44 in the Table.

12. *Discord with the extraneous intervals of a diminished minor 7th and a diminished minor 6th formed with the bass.*

Diminished minor 7th, diminished minor 6th, and minor 3rd $\begin{matrix} 7 \\ \flat 6 \end{matrix}$

No. 45 in the Table.

It is to be observed, that, in classing such discords as these (viz. discords with augmented or diminished intervals) as *extraneous*, that is, as presenting combinations not to be found in the scale, the term scale is to be understood in its strict sense as referring to the *proper* or *regular* series of sounds in the major and minor modes respectively. And this being understood, the classification is correct; for the discords in question all comprise combinations of this kind. Thus F, G♯, B, D (which is a chord of the major 6th, tritone, and augmented 2nd)*, cannot be found in the regular series of any

---

* Vide sup. p. 45.

key, either in the major or minor mode ; for none contains, in
its regular series*, the two notes F and G ♯.   And the case
is the same with respect to E, C, G ♯, which is a chord of the
minor 6th and major 3rd† ; for no key, major or minor, contains in its regular series‡ the two notes C and G ♯.   And so
of all other chords containing any augmented or diminished
interval.   But the term scale may, on the other hand, be taken
to include, not only the proper or regular series of sounds, in
either mode, but also the *augmented* (or *accidental*) notes of
the minor mode *ascending ;* and, in that case, the greater part
of the discords in question will be found to present *no* combination foreign to the scale, in that sense of the term.   Thus
all the notes of the above-named chords, F, G ♯, B, D, and
E, C, G ♯, may be found in the scale of A minor§, if both
its descending and ascending scale, including the augmented or accidental notes of the latter, be taken into the
account.   Such discords as these therefore, viz. such as are
produced by the aid of the augmented notes in the minor
mode, are capable of being considered either as extraneous
to the scale generally, or as belonging to the scale in the
minor mode, and extraneous to it in the major only ; and it is
in this latter light that they are in fact usually regarded in the
treatises.   But the former view of them seems better adapted
to the purposes of scientific arrangement, and the rather,
because *others* of the discords, with augmented or diminished
intervals, are strictly and absolutely extraneous ; so as not to
be found in any key, either in the major or minor mode, even
though the augmentations of the latter be taken in account.
And among these is the chord of the *extreme sharp* 6th‖,
which, as will appear hereafter, is a chord of importance.

Many of the extraneous discords are introduced occasionally, and some few of them copiously, into harmony
(more particularly by the modern masters), and often with

---

* Vide the Synopsis, sup. pp. 15, 16.   † Vide sup. p. 46.
‡ Vide Synopsis.   § Vide the Synopsis, p. 16.
‖ Vide sup. p. 46.

49

the greatest advantage ; for as discords prevent the monotony of incessant concords, so do extraneous discords the constant iteration of those proper to the scale*. In the subjoined Table, examples of the principal chords of this kind, according to the enumeration of them already given, will be found. Among them, the chord of the diminished 7th is, beyond comparison, the most common and the most important ; and the next rank is clearly due to the chord of the extreme sharp 6th, which is second indeed to none in felicity of effect.

Having now considered, separately, both divisions of discords, viz. such as belong to the scale, and such as are extraneous, we may next remark, in reference to the former, that the greater part of them most commonly, and in reference to to the latter, that some of them occasionally, occur in the shape of *chords of suspension*. A chord of suspension is any chord which suspends, for the moment, the harmony of some other familiar chord about to be played (particularly the common chord and dominant 7th and their inversions) ; which is effected by employing, in the chord of suspension, one or more of the notes of the chord to be suspended, but omitting some other or others of them, and retaining, in lieu of these, some note or notes of the chord immediately preceding. A chord of suspension is attended with this particular effect—that it gives piquancy to the chord suspended ; for, by interposing, just before it is sounded, a note or notes foreign to its proper harmony, it causes the ear to receive that harmony, when immediately afterwards heard, with the greater relish. It is to be observed, that chords of suspension must always fall upon an *accented* part of the bar, and the chords they suspend, upon an *unaccented* part. Many examples of them will be found in the subjoined Table of Chords ; for instance, in No. 7, where the chord of $\frac{5}{4}$, C, G, C, F, on the accented part of the bar, is a chord of suspension, suspending the harmony of the common chord that

* See Appendix, Note (BB).

E

follows on the unaccented part; which it does by employing only two of its notes (C and G), retaining at the same time the note F out of the first chord, B, G, D, F.

We have thus endeavoured to give some account of the different chords, whether concords or discords, which occur in music. For the more perfect elucidation of the subject, however, the following remarks must be added.

First. The account above given uniformly supposes the chords to be in their *perfect state*. In fact, however, they very frequently (and indeed *most* frequently) appear in a *thinned* or reduced state; for any of the notes of a chord (the bass excepted)* may, in general, be omitted without essential alteration of its character. Thus the common chord may present itself in the shape of $\frac{8}{3}$, instead of $\begin{smallmatrix}8\\5\\3\end{smallmatrix}$, the 5th being omitted; and the chord, $\begin{smallmatrix}6\\4\\2\end{smallmatrix}$, in the shape of $\frac{4}{2}$, the 6th being omitted.

Secondly. Though the different chords that have been enumerated, as well as some others not noticed, are all of occasional occurrence in music, they differ very greatly from each other as to the degree of frequency with which they occur, and the degree of importance which belongs to them. There are two, in particular, which, whether considered in their direct or their inverted form, are incomparably superior in these respects to the rest; viz. the *common chord*, and that species of the chord of the 7th which occurs on the *dominant* of the key, and is consequently called the *chord of the dominant 7th*†. These chords indeed, with their inversions, constitute, taken together, the greater portion of the whole system of harmony.

The subjoined Table exhibits the principal chords in one combined view, and illustrates them by Examples.

---

* See Appendix, Note (CC).   † Vide sup. p. 43.

EXAMPLES.

No. 1.

(In three different positions, according as the 8th, 3rd, or 5th, is uppermost.)*

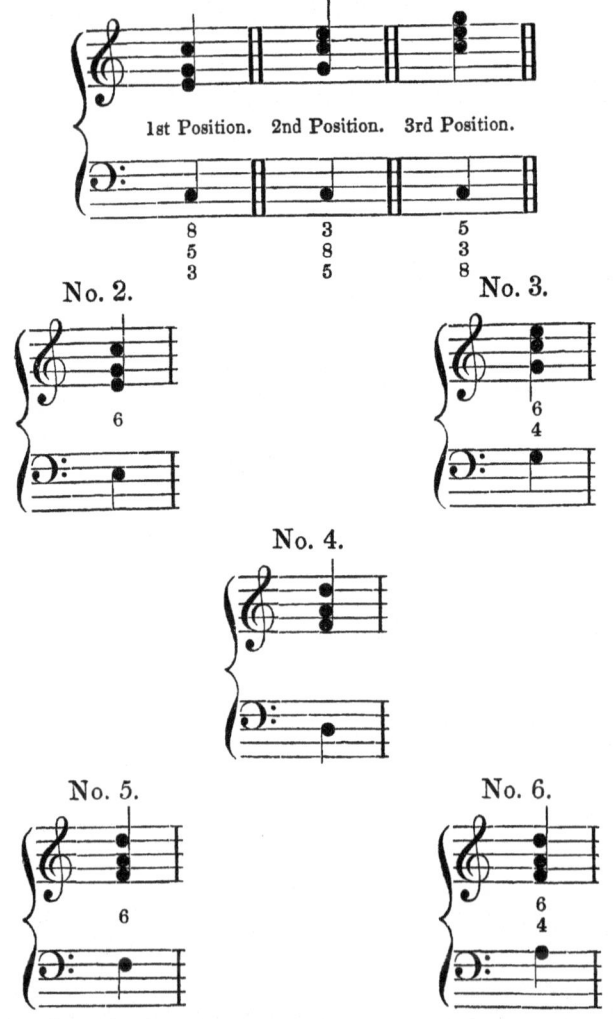

1st Position.   2nd Position.   3rd Position.

No. 2.   No. 3.

No. 4.

No. 5.   No. 6.

* It is to be observed, that the uppermost notes here, are, properly speaking, a 15th, 17th, and 19th from the bass note; though above described as an 8th, 3rd, and 5th. *Compound* intervals are constantly mentioned in this way, as if they were *single*.

TABLE OF THE PRINCIPAL CHORDS.

CONCORDS.

No. 1.

COMMON CHORD.

Figured $\begin{smallmatrix} 8 \\ 5 \\ 3 \end{smallmatrix}$, $\begin{smallmatrix} 8 \\ 3 \end{smallmatrix}$, 8, 5, or 3 (or without figure), consisting of an 8th, a 5th, and a major 3rd.

No. 2.	No. 3.
FIRST INVERSION	SECOND INVERSION
OF THE SAME	OF THE SAME
(viz. 3rd in the bass) called	(viz. 5th in the bass), called
chord of the 6th, figured 6, $\begin{smallmatrix} 8 \\ 3 \end{smallmatrix}$	chord of the $\begin{smallmatrix} 6 \\ 4 \end{smallmatrix}$, figured 6, $\begin{smallmatrix} 8 \\ 4 \end{smallmatrix}$
$\begin{smallmatrix} 6 \\ 3 \end{smallmatrix}$, or 6, consisting of an 8th, a minor 6th, and a minor 3rd.	or $\begin{smallmatrix} 6 \\ 4 \end{smallmatrix}$, consisting of an 8th, a major 6th, and a 4th.

No. 4.

COMMON CHORD MINOR.

Same as No. 1, sup. (except that it has a minor 3rd instead of a major), and figured as No. 1.

No. 5.	No. 6.
FIRST INVERSION	SECOND INVERSION
OF THE SAME	OF THE SAME
(viz. 3rd in the bass), called and figured as No. 2, sup. consisting of an 8th, a major 6th, and major 3rd.	(viz. 5th in the bass), called and figured as No. 3, sup. consisting of an 8th, a minor 6th, and a 4th.

No. 7.

No. 8.

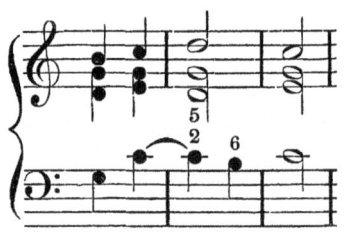

No. 9.

No. 10.

DISCORDS*.

I. DISCORDS OF THE SCALE.

No. 7.

$$\text{CHORD OF THE } \begin{matrix} 8 \\ 5 \\ 4 \end{matrix}$$

Figured $\begin{smallmatrix} 8 \\ 5 \\ 4 \end{smallmatrix}$, $\begin{smallmatrix} 5 \\ 4 \end{smallmatrix}$, or 4, consisting of an 8th, a 5th, and a 4th.

No. 8.

$$\text{CHORD OF THE } \begin{matrix} 5 \\ 2 \end{matrix}.$$

Figured $\begin{smallmatrix} 5 \\ 2 \end{smallmatrix}$, consisting of a 5th and a 2nd.

No. 9.

$$\text{CHORD OF THE } \begin{matrix} 5 \\ 4 \\ 2 \end{matrix}$$

Figured $\begin{smallmatrix} 5 \\ 4 \\ 2 \end{smallmatrix}$, consisting of a 5th, a 4th, and a 2nd.

No. 10.

IMPERFECT COMMON CHORD.

Occurring only upon the 7th (or leading note) of the key (if the mode be major), figured as No. 1, sup. (or without figure), consisting of an 8th, a minor 5th, and a minor 3rd.

---

* The discords are arranged according to the Schemes, sup. pp. 41, 45, with only this exception, that those schemes rank every chord as an independent chord ; but the Table ranks some of them as *inversions* of others, in which light these particular chords are always most conveniently considered in practice. These chords are Nos. 12, 13, 14, 16, 17, 18, 41, 42, 43.

No. 11.

(In three different positions, according as the 3rd, 5th, or 7th is uppermost.)

No. 12.	No. 13.	No. 14.

No. 15.

No. 16.	No. 17.	No. 18.

No. 11.

Chord of the Dominant 7th.

Occurring only upon the dominant of the key, figured $\begin{smallmatrix}7\\5\\3\end{smallmatrix}$, $\begin{smallmatrix}7\\5\end{smallmatrix}$, 7, consisting of a minor 7th, 5th, and major 3rd.

No. 12.

First Inversion of the same

(viz. 3rd in the bass) called chord of the $\begin{smallmatrix}6\\5\end{smallmatrix}$, figured $\begin{smallmatrix}6\\5\\3\end{smallmatrix}$, $\begin{smallmatrix}6\\5\end{smallmatrix}$, consisting of minor 6th, minor 5th, and minor 3rd.

No. 13

Second Inversion of the same

(viz. 5th in the bass) called chord of the $\begin{smallmatrix}6\\4\\3\end{smallmatrix}$, or $\begin{smallmatrix}4\\3\end{smallmatrix}$, and so figured, consisting of major 6th, 4th, and minor 3rd.

No. 14.

Third Inversion of the same

(viz. 7th in the bass) called chord of the $\begin{smallmatrix}6\\4\\2\end{smallmatrix}$, $\begin{smallmatrix}4\\2\end{smallmatrix}$, or 2, and so figured, consisting of major 6th, tritone, and 2nd.

No. 15.

Chord of the 7th upon the 2nd of the Key.

Occurring only on the 2nd of the key (if the mode be major), figured as No. 11, consisting of a minor 7th, 5th, and minor 3rd.

No. 16.

First Inversion of the same

(viz. 3rd in the bass) called and figured as No. 12, sup. consisting of major 6th, 5th, and major 3rd.

No. 17.

Second Inversion of the same

(viz. 5th in the bass) called and figured as No. 13, sup. consisting of minor 6th, 4th, and minor 3rd.

No. 18.

Third Inversion of the same

(viz. 7th in the bass) called and figured as No. 14, sup. consisting of major 6th, 4th, and 2nd.

No. 19.

No. 20.

No. 21.

No. 22.

No. 19.

$$\text{CHORD OF THE } \begin{smallmatrix} 7 \\ 4 \\ 3 \end{smallmatrix}.$$

Figured $\begin{smallmatrix} 7 \\ 4 \\ 3 \end{smallmatrix}$, or $\begin{smallmatrix} 7 \\ 4 \end{smallmatrix}$, consisting of a minor 7th, 4th, and minor 3rd.

No. 20.

$$\text{CHORD OF THE } \begin{smallmatrix} 7 \\ 6 \\ 3 \end{smallmatrix}.$$

Figured $\begin{smallmatrix} 7 \\ 6 \\ 3 \end{smallmatrix}$, or $\begin{smallmatrix} 7 \\ 6 \end{smallmatrix}$, consisting of a minor 7th, major 6th, and major 3rd.

No. 21

$$\text{CHORD OF THE } \begin{smallmatrix} 7 \\ 5 \\ 4 \end{smallmatrix}.$$

Figured $\begin{smallmatrix} 7 \\ 5 \\ 4 \end{smallmatrix}$, consisting of a minor 7th, 5th, and 4th.

No. 22.

$$\text{CHORD OF THE } \begin{smallmatrix} 7 \\ 4 \\ 2 \end{smallmatrix}.$$

Figured $\begin{smallmatrix} 7 \\ 4 \\ 2 \end{smallmatrix}$, consisting of a major 7th, 4th, and 2nd.

No. 23.

No. 24.

No. 25.

No. 26.

No. 23.

$$\text{CHORD OF THE } \begin{matrix} 7 \\ 6 \\ 4 \\ 2 \end{matrix}.$$

Figured $\begin{matrix} 7 \\ 6 \\ 4 \\ 2 \end{matrix}$, consisting of a major 7th, a major 6th, a 4th, and a 2nd.

No. 24.

$$\text{CHORD OF THE } \begin{matrix} 7 \\ 5 \\ 4 \\ 2 \end{matrix}.$$

Figured $\begin{matrix} 7 \\ 5 \\ 4 \\ 2 \end{matrix}$, consisting of a major 7th, a 5th, a 4th, and a 2nd.

No. 25.

CHORD OF THE 9TH.

Figured $\begin{matrix} 9 \\ 5 \\ 3 \end{matrix}$, 9, consisting of a major 9th, a 5th, and a major 3rd.

No. 26.

$$\text{CHORD OF THE } \begin{matrix} 9 \\ 6 \\ 3 \end{matrix}.$$

Figured $\begin{matrix} 9 \\ 6 \\ 3 \end{matrix}$, $\begin{matrix} 9 \\ 6 \end{matrix}$, consisting of a minor 9th, a minor 6th, and a minor 3rd.

No. 27.

No. 28.

No. 29.

No. 30.

No. 27.

CHORD OF THE $\frac{9}{6}$.
4

Figured $\frac{9}{6}$, consisting of a minor 9th, a minor 6th, and a 4th.
4

No. 28.

CHORD OF THE $\frac{9}{5}$.
4

Figured $\frac{9}{5}$, $\frac{9}{4}$, consisting of a major 9th, a 5th, and a 4th.
4

No. 29

CHORD OF THE $\frac{9}{7}$.
3

Figured $\frac{9}{7}$, consisting of a major 9th, a minor 7th, and a
3, minor 3rd.

No. 30.

CHORD OF THE $\frac{9}{7}$.
4

Figured $\frac{9}{7}$, consisting of a major 9th, a major 7th, and a 4th.
4,

No. 31.

No. 32.

No. 33.

No. 34.

No. 35.

II. EXTRANEOUS DISCORDS.

No. 31.

CHORD OF THE MAJOR 7TH AND AUGMENTED 2ND.

Figured $\frac{7}{2}$, consisting of a major 7th and an augmented 2nd.

No. 32.

CHORD OF THE MINOR 6TH AND DIMINISHED 4TH.

Figured $\frac{6}{b\,4}$, consisting of a minor 6th and diminished 4th.

No. 33.

CHORD OF THE MINOR 6TH AND MAJOR 3RD.

Figured 6, consisting of a minor 6th and major 3rd.

No. 34.

COMMON CHORD WITH AUGMENTED 5TH.

Figured 5, consisting of an 8th, an augmented 5th, and a major 3rd.

No. 35.

CHORD OF THE MINOR 7TH, MINOR 5TH, AND MAJOR 3RD.

Figured $\frac{7}{5}$, consisting of a minor 7th, a minor 5th, and a major 3rd.

F

No. 36.

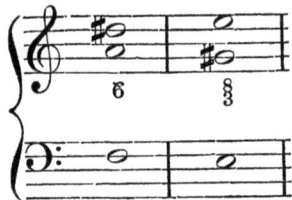

No. 37.

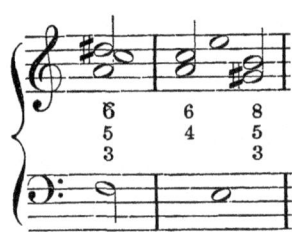

No. 38.

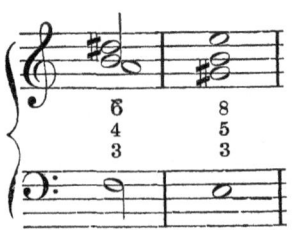

No. 39.

No. 36.

Chord of the Augmented Major 6th (or Extreme sharp 6th).

Figured 6, consisting of an augmented major 6th and a major 3rd.

No. 37.

Chord of the Augmented Major 6th with a 5th.

Figured $\begin{smallmatrix} 6 \\ 5 \\ 3 \end{smallmatrix}$, consisting of an augmented major 6th, a 5th, and a major 3rd.

No. 38.

Chord of the Augmented Major 6th with a Tritone.

Figured $\begin{smallmatrix} 6 \\ 4 \\ 3 \end{smallmatrix}$, consisting of an augmented major 6th, a tritone, and a major 3rd.

No. 39.

Chord of the Major 7th, Minor 6th, 4th, and 2nd.

Figured $\begin{smallmatrix} 7 \\ 6 \\ 4 \\ 2 \end{smallmatrix}$, consisting of a major 7th, a minor 6th, a 4th, and a 2nd.

No. 40.

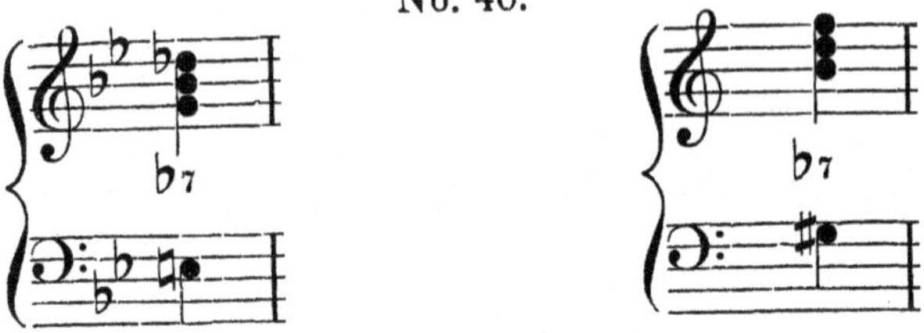

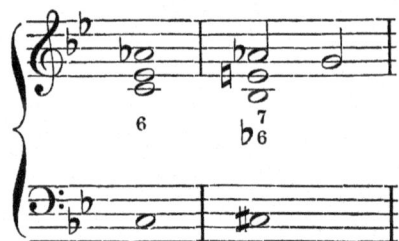

No. 41.    No. 42.    No. 43.

No. 44.

No. 45.

No. 40.

CHORD OF THE DIMINISHED 7TH.

Figured 7 or ♭7 with a ♮ or ♯ at the bass note ; consisting of a diminished minor 7th, a diminished 5th, and a minor 3rd.

No. 41.	No. 42.	No. 43.
FIRST INVERSION (viz. 3rd in the bass) called and figured 6 chord of the 5, con-3 sisting of a major 6th, a minor 5th, and a minor 3rd.	SECOND INVERSION (viz. 5th in the bass) called and figured 6 chord of the 4, con-3 sisting of a major 6th, an augmented 4th, and a minor 3rd.	THIRD INVERSION (viz. 7th in the bass) called and figured 6 chord of the 4, con-2 sisting of a major 6th, a tritone, and an augmented 2nd.

No. 44.

CHORD OF THE DIMINISHED MINOR 7TH, DIMINISHED 5TH, AND DIMINISHED MINOR 3RD.

Figured $\frac{7}{♭5}$, consisting of a diminished minor 7th, a diminished 5th, and a diminished minor 3rd.

No. 45.

CHORD OF THE DIMINISHED MINOR 7TH, DIMINISHED MINOR 6TH, AND MINOR 3RD.

Figured $\frac{7}{♭6}$, consisting of a diminished minor 7th, a diminished minor 6th, and a minor third.

CHAPTER III.

A succession of chords may be put as an accompaniment
to a given melody (or air), by placing under each note to be
accompanied some one of its primary fundamental basses*, and
making such fundamental bass bear its common chord major;
including, among the constituent notes of that chord, the note
of the melody which is intended to be accompanied.   But as
such an accompaniment would contain a series of bass notes
(or, as it is more shortly termed, a *bass*), consisting exclusively of
the three primary fundamental basses of the key, and the
harmonies would be all those of the common chord major,
the effect would be monotonous and uninteresting; and therefore such method is only pursued in part, and recourse is
had to other expedients to give more variety and force to the
accompaniment.   For, first, the common chords major are very
frequently taken in their *inverted*, instead of their *direct* form.
Next resort is often had to some of the *secondary* fundamental basses, and to their common chords *minor*, direct or inverted.   And, lastly, very frequent recourse is made (both

---

* Vide sup. p. 36.

for the sake of *variety* and *contrast*) to *discords*, direct and inverted, and to changes of key (called *modulations*) ; and even to *extraneous discords* ; and these also are taken both in their direct and inverted form*.

As regards the choice between the different fundamental basses (primary or secondary) capable of being assigned to the same note of the melody, it is to be observed, first, that the primary fundamental basses are in general employed in preference to the secondary ones, because it is by the former that the key and mode are more particularly characterized† ; and, secondly, that in passing from one common chord major, minor, or imperfect (or its inversion), to another common chord major, minor, or imperfect (or its inversion), it is a general rule (though subject to frequent exception, according to the particular effect that may be intended) that the fundamental basses should succeed each other by concordant intervals, rather than by seconds or sevenths ; the reason for which is a clear and satisfactory one ; viz. that, where the progression is by seconds and sevenths, it results from the very constitution of the scale, that there can be no *liaison* or connection between the common chords (or inversions) employed ; that is, they can have no note in common‡, as, by a rule to be hereafter noticed, they ought in general to have, and as they always will have, where the progression of the fundamental bass is by consonant intervals. The concordant intervals most regularly applicable to the purpose, are that of an ascending fourth (equivalent to a descending fifth), and that of an ascending sixth (equivalent to a descending third); and of these, the former is generally preferable. But the fundamental bass sometimes proceeds by an ascending fifth (equivalent to a descending fourth); viz. in passing from the common chord of the subdominant to the common chord of the tonic, and from the common chord of the tonic to the common chord of the

---

* See Appendix, Note (DD).   † Vide sup p. 36.

‡ See Appendix, Note (EE)·

dominant. In some cases, also, it proceeds by an ascending third (equivalent to a descending sixth), or even by an ascending second or an ascending seventh, equivalent to a descending seventh and a descending second respectively. The progression, however, by a second or a seventh is somewhat harsh; and when it occurs (as it frequently does), it is common to make the melody and bass proceed by *contrary motion ;* that is, one of them is made to *ascend,* and the other to *descend ;* as exemplified in the ascending and descending scale in the next page*; it being found that this method has the effect of reconciling the ear to the progression.

The chords, of which the melody and its accompaniment consist, are usually taken in such *positions*† that the notes of the melody constitute in each case the *uppermost* notes (or *parts*) of the chords ; but this is not invariably the case—the melody being sometimes placed among the lower parts.

The lowest part, or bass, is the part on which the character of the whole accompaniment chiefly depends ; and therefore the composer must be careful, in the first place, to assign to this part such a series of notes as shall be most suitable to the melody ; and thus, paying his principal attention to the combination of these parts, viz. the melody and the bass, he will be led to make the other parts in some degree subservient. With respect to the bass, we may also take occasion to remark, that its concluding note is regularly always the tonic, accompanied by its common chord; or (in the minor mode) its common chord minor.

We shall now give examples of the manner in which a melody is accompanied, taking in the first instance the scale itself for a melody‡ ; and shall give it an accompaniment consisting exclusively of common chords major, direct or inverted, or, in other words, founded exclusively on the primary fundamental basses. After which, we shall exhibit the scale with an accompaniment constructed in the ordinary and more

---

* See the 6th and 7th chords ascending, and the 2nd and 3rd descending.

† Vide sup. p. 32.   ‡ See Appendix, Note (FF).

---

effective manner, with an intermixture of common chords, major and minor, and their inversions, and of discords direct and inverted, and with a modulation. And we shall subjoin, in each case, to the accompaniment (in letters), the fundamental basses, primary and secondary, on which it is founded; and add an analysis, explaining the course of the harmony, and shewing that it proceeds upon the principles which are above stated in reference to the method pursued in putting an accompaniment to a melody.

THE SCALE, WITH AN ACCOMPANIMENT OF COMMON CHORDS MAJOR, DIRECT AND INVERTED.

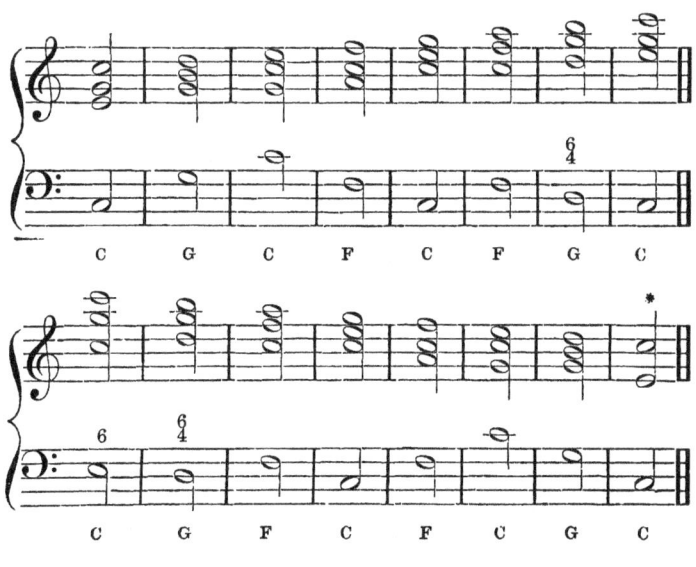

* It is to be observed that this chord, though apparently consisting of only three notes, consists, in reality (like the rest), of four. For, it is to be considered as C E C C; so that, if sounded by four voices and instruments, two of them would sound C and E, and the other two C and C. When the same note (as C in this case) is thus taken by different voices or instruments, the sounds are described as *unisons*.

ANALYSIS.

In the first bar,

The tonic (in the key of C) is accompanied by the common chord major of the tonic.

In the second bar,

The second, by the ditto of the dominant.

In the third bar,

The third, by the ditto of the tonic.

In the fourth bar,

The subdominant, by the ditto of the subdominant.

In the fith bar,

The dominant, by the ditto of the tonic.

In the sixth bar,

The sixth, by the ditto of the subdominant.

In the seventh bar,

The leading note, by the second inversion of the chord of the dominant.

In the eighth bar,

The octave of the tonic, by the common chord major of the tonic.

In the ninth bar,

The octave of the tonic, by the first inversion of the common chord of the tonic.

In the tenth bar,

The leading note, by the second inversion of the common chord major of the dominant.

In the eleventh bar,

The sixth, by the common chord major of the subdominant.

In the twelfth bar,

The dominant, by the ditto of the tonic.

In the thirteenth bar,

The subdominant, by the ditto of the subdominant.

In the fourteenth bar,
The third, by the ditto of the tonic.
In the fifteenth bar,
The second, by the ditto of the dominant.
In the sixteenth bar,
The tonic by the ditto of the tonic.

THE SCALE WITH AN ACCOMPANIMENT OF COMMON CHORDS
MAJOR AND MINOR, DIRECT AND INVERTED, AND DISCORDS,
DIRECT AND INVERTED, AND WITH A MODULATION.

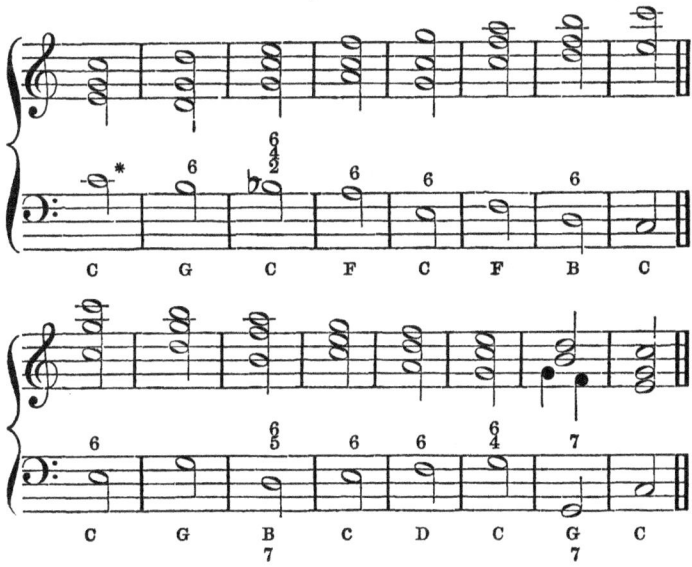

ANALYSIS.

In the first bar,
The tonic in the key of C is accompanied by the common chord major of the tonic.

In the second bar,
The second, by the first inversion of the common chord major of the dominant.

---

* In figuring chords, the common chords are, in general, not figured.

In the third bar,

The leading note in the key of F, by the third inversion of the chord of the dominant seventh.

In the fourth bar,

The tonic in the key of F, by the first inversion of the common chord of the tonic.

In the fifth bar,

The second in the key of F, by the first inversion of the common chord of the dominant.

In the sixth bar,

The third in the key of F, by the common chord of the tonic.

In the seventh bar,

The leading note, in the original key of C, by the first inversion of the imperfect common chord.

In the eighth bar,

The octave of the tonic, by the common chord of the tonic.

In the ninth bar,

The octave of the tonic, by the first inversion of the common chord of the tonic.

In the tenth bar,

The leading note, by the common chord of the dominant.

In the eleventh bar,

The sixth, by the first inversion of the chord of the seventh on the leading note.

In the twelfth bar,

The dominant, by the first inversion of the common chord major of the tonic.

In the thirteenth bar,

The subdominant, by the first inversion of the common chord minor of the second.

In the fourteenth bar,
The third, by the second inversion of the common chord of the tonic.

In.the fifteenth bar,
The second, by the chord of the dominant seventh.

In the sixteenth bar,
The tonic, by the common chord of the tonic.

We shall now proceed to examine, in the same manner, some well-known psalm tunes, this being a species of music very suitable for the illustration of the elementary principles of accompaniment.

PSALM TUNE.

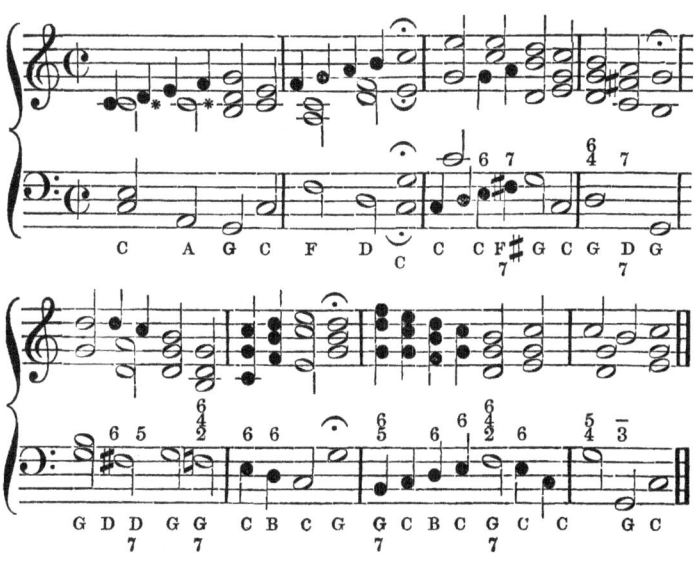

---

* Notes like the D and F here, passing, by the interval of a second, from the notes of a preceding concord, and found on a *weak* part of the bar (vide sup. p. 22) are called *passing* notes, and are not considered as essential parts of the composition. Even where figures are used, therefore, such notes are in general without figures, as in the case before noticed of common chords.

ANALYSIS.

The notes of the melody, with the accompanying chords, are as follows:

In the first bar.

The tonic, in the key of C, is accompanied by the common chord of the tonic.

The third, by the common chord minor of the 6th.

The dominant, by the common chord of the dominant.

The third, by the common chord of the tonic.

In the second bar,

The subdominant, by the common chord of the subdominant.

The sixth, by the common chord minor of the second.

The tonic, by the common chord of the tonic.

In the third bar,

The third, by the common chord of the tonic.

The third, by the first inversion of the common chord of the tonic, and by the chord of the seventh on the leading note in the key of G.

The dominant in the key of G, by the common chord of the tonic.

The subdominant in the key of G, by the common chord of the subdominant.

In the fourth bar,

The third in the key of G, by the second inversion of the common chord of the tonic.

The second in the key of G, by the chord of the dominant seventh.

The tonic in the key of G, by the common chord of the tonic.

In the fifth bar,

The dominant in the key of G, by the common chord of the tonic.

The dominant in the key of G, by the first inversion of the common chord of the dominant.

The subdominant in the key of G, by the first inversion of the chord of the dominant seventh.

The third, in the key of G, by the common chord of the tonic.

The dominant in the original key of C, by the third inversion of the chord of the dominant seventh.

In the sixth bar,

The tonic, by the first inversion of the common chord of the tonic.

The second, by the first inversion of the imperfect common chord.

The third, by the common chord of the tonic.

The second, by the common chord of the dominant.

In the seventh bar,

The subdominant, by the first inversion of the chord of the dominant seventh.

The third, by the common chord of the tonic.

The second, by the first inversion of the imperfect common chord.

The tonic, by the first inversion of the common chord of the tonic.

The leading note, by the third inversion of the chord of the dominant seventh.

The tonic, by the first inversion of the common chord of the tonic, and by the common chord of the tonic.

In the eighth bar,

The tonic, by the chord of $\frac{5}{4}$ on the dominant, suspending the common chord of the dominant.

The leading note, by the common chord of the dominant.

The tonic, by the common chord of the tonic.

PSALM TUNE.

ANALYSIS.

The notes of the melody, with the accompanying chords, are as follows :

In the first bar,

The tonic, in the key of C, is accompanied by the common chord of the tonic.

In the second bar,

The ditto ——————————— by ditto ——————————————

The leading note, by the first inversion of the chord of the seventh upon the tonic.

The sixth, by the common chord of the subdominant.

The dominant, by the common chord of the dominant.

In the third bar,

The dominant, by the common chord of the tonic.

The subdominant, by the first inversion of the imperfect common chord.

The third, by the first inversion of the common chord of the tonic.

In the fourth bar.

The sixth, by the common chord of the subdominant.

The dominant, by the first inversion of the common chord of the tonic.

The sixth, by the common chord of the subdominant.

The leading note, by the chord of the dominant seventh.

In the fifth bar,

The tonic, by the common chord of the tonic

In the sixth bar,

The leading note, by the common chord of the dominant.

The second, by the common chord of the dominant.

In the seventh bar,

The second, by the common chord of the dominant.

The ditto, by ditto.

In the eighth bar,

The third, by the common chord of the tonic.

Ditto,   by ditto.

In the ninth bar,

The second, by the common chord of the dominant.

The third, by the common chord of the tonic.

The second, by the common chord minor of the second.

G

In the tenth bar,

The tonic, by the second inversion of the common chord minor of the sixth.

The second, in the key of A minor, by the chord of the dominant seventh.

In the eleventh bar,

The tonic in the key of A minor, by the common chord minor of the tonic.

In the twelfth bar,

The dominant in the original key of C, by the common chord of the tonic.

The third, by the common chord of the tonic.

The dominant, by the first inversion of the common chord of the tonic.

In the thirteenth bar,

The tonic, by the first inversion of the chord of the seventh upon the second.

The leading note, by the common chord of the dominant.

In the fourteenth bar,

The sixth, by the first inversion of the common chord minor of the second.

The tonic, by the common chord of the tonic.

The leading note, by the common chord of the dominant.

The sixth, by the first inversion of the common chord minor of the second.

In the fifteenth bar,

The dominant, by the chord of the 7th upon the 3d; suspending the first inversion of the common chord of the tonic.

The sixth, by the common chord of the subdominant.

The leading note, by the second inversion of the imperfect common chord.

The tonic, by the first inversion of the common chord of the tonic.

In the sixteenth bar,

The second, by the common chord minor of the second.

The tonic, by the chord of the seventh on the second.

The leading note, by the common chord of the dominant, and by the chord of the dominant seventh.

In the seventeenth bar,

The tonic, by the common chord of the tonic.

PSALM TUNE.

Harrington.

G 2

ANALYSIS.

The notes of the melody, with the accompanying chords, are as follows:

In the first bar,

The dominant in the key of E♭, is accompanied by the common chord of the tonic.

In the second bar,

The tonic, by the first inversion of the common chord of the tonic.

The leading note, by the first inversion of the imperfect common chord.

The sixth, by the first inversion of the common chord minor of the 6th.

In the third bar,

The dominant, by the first inversion of the common chord of the dominant.

The subdominant, by the chord of the dominant seventh.

The third, by the common chord of the tonic.

In the fourth bar,

The subdominant, by the first inversion of the common chord of the subdominant.

The dominant, by the second inversion of the common chord of the tonic.

The sixth, by the common chord of the subdominant.

In the fifth bar,

The dominant, by the common chord of the tonic.

Ditto, by the first inversion of the common chord of the dominant.

In the sixth bar,

The dominant, by the chord of $\frac{7}{4}$ on the 6th.
3

The leading note in the key of B♭, by the first inversion of the imperfect common chord.

The tonic in the key of B♭, by the common chord of the tonic.

In the seventh bar,

The sixth in the key of B♭, by the common chord of the subdominant.

The second in the key of B♭, by the first inversion of the common chord minor of the second.

The tonic in the key of B♭, by the second inversion of the common chord of the tonic.

The leading note in the key of B♭, by the chord of the dominant 7th.

In the eighth bar,

The tonic in the key of B♭, by the common chord of the tonic.

In the ninth bar.

The dominant in the key of B♭, by the common chord of the tonic.

The tonic in the key of B♭, by the first inversion of the common chord minor of the 6th.

In the tenth bar,

The tonic in the key of B♭, by the first inversion of the common chord minor of the 6th.

The subdominant in the original key of E♭, by the chord of the dominant 7th.

The third, by the common chord of the tonic.

In the eleventh bar,

The third, by the second inversion of the common chord of the tonic.

The second, by the third inversion of the chord of the dominant 7th.

The tonic, by the first inversion of the common chord of the tonic.

In the twelfth bar,

The second, by the second inversion of the common chord of the dominant.

The subdominant, by the imperfect common chord.

The third, by the common chord of the tonic.

In the thirteenth bar,

The sixth, by the common chord of the subdominant.

The dominant, by the first inversion of the common chord of the tonic.

The tonic, by ditto.

In the fourteenth bar,

The tonic, by the chord of $\frac{7}{4}{3}$ on the second.

The leading note, by the chord of the dominant 7th.

The tonic, by the common chord of the tonic.

The third, by ditto.

In the fifteenth bar,

The second, by the first inversion of the chord of the seventh on the second.

The third, by the second inversion of the common chord of the tonic.

The second, by the common chord of the dominant, and by the chord of the dominant 7th.

In the sixteenth bar,

The tonic, by the common chord of the tonic.

These Examples, it will be observed, are all in the major mode. But it may be useful to present the reader with a sample of the *minor* also*; and for this purpose we will select the following

CHANT.

G   G   D   G

---

* Appendix (GG)

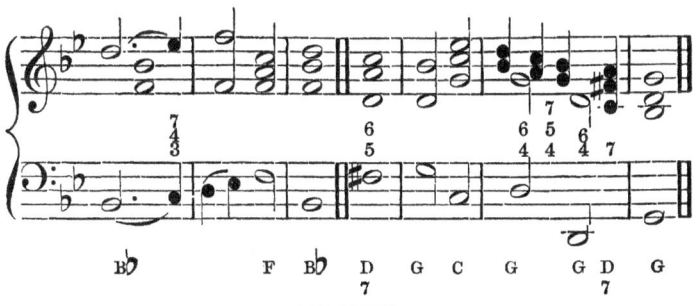

ANALYSIS.

The notes of the melody, with the accompanying chords, are as follows :—

In the first bar,

The third ascending in G minor, by the common chord minor of the tonic. The subdominant, by the chord of $\frac{7}{4}$ on the second ascending.

In the second bar,

The dominant, by the first inversion of the common chord minor of the tonic.

The second ascending, by the common chord of the dominant.

In the third bar,

The third ascending by the common minor of the tonic.

In the fourth bar,

The dominant, by its octave below.

In the fifth bar,

The sixth ascending and the dominant, by their octaves below.

In the sixth bar,

The tonic, second, third, and subdominant, by their octaves below.

In the seventh bar,

The dominant, by its octave below.

In the eighth bar,

The dominant, by the common chord major of the third

ascending. The third descending, by the chord of $\frac{7}{4}$ on the subdominant. $\quad 3$

In the ninth bar,

The second descending, by the fourth descending.

The subdominant, by the common chord major of the second descending.

In the tenth bar,

The dominant, by the common chord major of the third ascending.

In the eleventh bar,

The subdominant, by the first inversion of the chord of the dominant seventh.

In the twelfth bar,

The third ascending, by the common chord minor of the tonic.

The third descending, by the common chord minor of the subdominant.

In the thirteenth bar,

The dominant, by the second inversion of the common chord minor of the tonic.

The sub-dominant, by the chord of $\frac{7}{5}$ on the dominant. $\quad 4$

The third ascending, by the second inversion of the common chord minor of the tonic.

The second ascending, by the chord of the dominant seventh.

In the fourteenth bar,

The tonic, by the common chord minor of the tonic.

In putting an accompaniment of chords to a melody (of which we have, hitherto, considered only the general method), the following Rules are also to be observed :—

I.   *Consecutive fifths* or octaves, by similar motion, are not allowable.*

The meaning of which is, that when any two parts in any

---

* See Appendix, Note (HH).

chord have made a 5th with each other, the same parts should not again immediately make a 5th with each other, in similar motion, whether ascending or descending; and that, when two parts in any chord have made an octave with each other, the same parts should not again immediately make an octave with each other, in similar motion, whether ascending or descending.

Thus:

Consecutive 5ths.

Ex. 1.

Albrechtsberger.

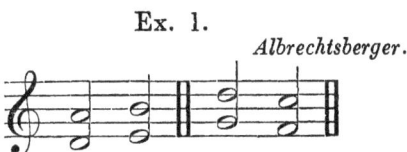

Ex. 2.

Corfe.

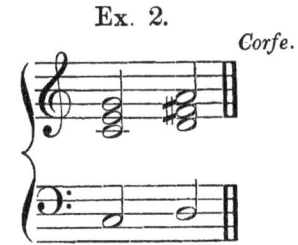

Ex. 3.

Shield.

Consecutive 8ths.

Ex. 1.

Albrechtsberger.

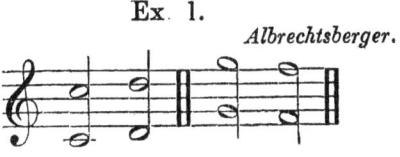

Ex. 2.

Corfe.

In the first of these examples of consecutive 5ths (ascending series), we may observe that there is a 5th between the bass part D, in the first chord, and the first part A ; and, again, a 5th between the bass part E, in the second chord, and the first part B ; and the parts of the second 5th have *similar motion*, that is, both ascend. So, in the descending series, there is a 5th between the bass part G, in the first chord, and the first part D ; and, again, a 5th between the bass part F, in the second chord, and the first part C ; and the parts of the second 5th have similar motion, that is, both descend. The intervals D—A and E—B, in the first case, are therefore consecutive 5ths forbidden by the Rule; and so are the intervals G—D and F—C, in the other.

So, in the second example of consecutive 5ths, there is a 5th between the bass part C, in the first chord, and the first part G; and, again, a 5th between the bass part D, in the second chord, and the first part A ; both of the parts of which latter 5th ascend. These intervals therefore, C—G and D—A, are faulty as consecutive 5ths.

In the third example, in like manner, each staff exhibits a succession of 5ths with similar motion; and though this difference may be remarked between them—that in the first staff the 5ths move *diatonically* (or by contiguous degrees), in the second by *skips*, yet they both alike offend against the rule under consideration.

Again, in the first example of consecutive 8ths, in the ascending series, there is an 8th between the bass part C, in the first chord, and the first part C; and an 8th between the bass part D, in the second chord, and the first part D; and the parts of the second 8th both ascend. Also, in the descending series, an 8th between the bass part G, in the first chord, and the first part G; and an 8th between the bass part F, in the second chord, and the first part F; and the parts of the second 8th both descend.

So, in the second example of consecutive 8ths, there is an 8th between the bass part D, in the first chord, and the first part D; and, again, an 8th with similar motion between the bass part C, in the second chord, and the first part C. This example also comprises the bad progression of consecutive 5ths; for, in the first chord, the bass part D makes a 5th with the second part A; and in the second chord, the bass part C again makes a 5th with the second part G; both parts descending.

Of these two faults of consecutive 5ths and consecutive 8ths, the latter is more tolerable to the ear than the former. The latter progression, indeed, sometimes takes place in certain passages by design, and may then be attended with a good effect.

Consecutive 5ths and 8ths, though bad by *similar motion*, are, on the other hand, allowable in the case of *contrary motion;* that is, where one of the parts of the second 5th or 8th is made to ascend, and the other to descend, or vice versâ. Thus:

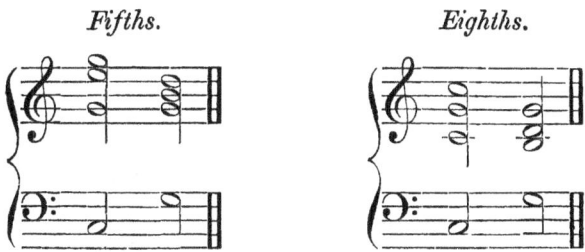

Fifths.　　　　　　*Eighths.*

For this is no violation of the rule; and by such a progression the ear is not found to be offended.

The subject of consecutive 5ths and 8ths may be further illustrated, by shewing how the fault may be *corrected*.

Consecutive 5ths and 8ths are both exhibited in the following example :

Corfe.

But the faults may be corrected as follows:

Ibid.

For in this case there is no immediate succession of 5ths between the same parts ; nor is there any succession of 8ths*.

II. *In general, every chord should have a connection (liaison) with the preceding one.*

By this rule, which is less distinctly laid down in the treatises than its important and fundamental character seems to require†, we mean that every chord (generally speaking) should contain one note or more in common with the preceding chord. The rule applies more particularly to the concords; for,

---

* See Appendix, Note (II).
† See Appendix, Note (KK).

though it be also generally true in regard to the discords, yet the connection of these with the preceding harmony is more specifically provided for by another rule, to be presently considered ; viz. that which requires discords (in general) to be *prepared*.

As regards the concords, it has been already shown that the rule now under consideration as to *liaison* (or, as it is also called by the French writers, *enchainement)* is enforced by a proper succession of the fundamental basses*.

Two successive chords are also in some degree connected with each other (but less perfectly) when the second contains a note differing by a semitone only from some note contained in the first. And this imperfect *liaison* sometimes occurs instead of the other.

III. *When both the bass and the treble parts move, they ought to take contrary motion as often as convenient.*

The nature of *contrary motion* has been already explained†; and we have seen that, by this method, the ear may be reconciled to the progression of the fundamental bass by seconds or sevenths ; and even to consecutive fifths or octaves. The present rule refers to its effect, as giving variety and point to harmony in general. Examples are of constant occurrence ; for the present purpose, it will be sufficient to refer to the psalm tune, sup. p. 77 ; where good illustrations of the progression by contrary motion occur.

IV. *Discords must be resolved ; and in some cases also prepared.*

A discord is said to be *resolved*, when it is followed in due progression by a concord ; and, in general, the progression

---

* Vide sup. p. 71.˙   † Vide. sup pp. 72, 91.

consists in making one of the notes which are in discord with each other *descend* a semitone (or sometimes a tone) in the next succeeding chord. Thus:

Or thus:

In the first of which Examples, the chord of the 9th, C E D, occurs, having the two discordant intervals, C and D, and E and D; and is resolved into the common chord (in the same bar), by D's descending a tone to C. And, in the next, the chord of the $\frac{9}{7}$, E D F, occurs, having the three discordant intervals, E and D, E and F, and G and F; and is resolved into a chord of the 6th, by F's descending a semitone to E, and D's descending a tone to C.

The manner of resolving a discord, however, if more particularly considered, depends upon the nature of the particular discord*. The most important one, viz. the chord of the dominant 7th (which has two discordant intervals, viz. that of the tritone or its inversion, the minor 5th, formed by the subdominant and leading note, and that of the minor 7th, formed

---

* See Appendix, Note (LL).

by the dominant and subdominant), is resolved by making the
leading note ascend a semitone, and the subdominant descend
a semitone (or, in the *minor* mode, a tone) ; and giving such
progressions to the other parts of the chord as shall be necessary to complete its conversion into the common chord (or, in
the minor mode, common chord minor) of the tonic.   Thus :

In the first of which examples, the chord of the dominant
7th, G D F B, is resolved into C C E C, the common chord
of the tonic C, by the ascent of the leading note B, a semitone
to C ; the descent of the subdominant F, a semitone to E ;
the descent of the 2nd D, a tone to C ; and the descent of
the dominant G, a 5th to C.   And in the second of which
examples, the chord of the dominant 7th, E B D G♯, is resolved into A A C A, the common chord minor of the tonic
A, by the ascent of the leading note, G♯, a semitone to A ; the
descent of the subdominant D, a tone to C ; the descent of the
2nd B, a tone to A ; and the descent of the dominant E, a
5th to A.
    The subject includes (it will be observed) the considera-
tion of the *time* at which the resolution of a discord may take
place ; for, as to this, too, there is a variety.   Though it

generally occurs in the next succeeding chord, it is sometimes deferred until several other chords have intervened.

In general, a discord must be *prepared*, as well as *resolved;* that is, one of the notes in discord with each other must be introduced as a concordant note in the chord immediately preceding. Thus, in the 1st of the Examples in p. 94, the discord of the $\frac{9}{8}$ is prepared, by the introduction of one of the discordant notes, D, into the previous chord in the character of a concordant note. And, in the 2nd Example, the discord of the $\frac{9}{7}$ is prepared by the introduction of F, one of the discordant notes, into the previous chord, in the like character. The effect of this preparation is to mitigate the harshness of the discord that follows. But there are discords which (owing, as it would seem, to their greater agreeableness to the ear*) require no preparation ; particularly the chords of the dominant 7th, and diminished 7th, and their inversions.

V. *As a general rule, discordant notes must not be doubled·*

By a *discordant note* is meant any note in a chord, by the introduction of which the whole chord (otherwise a concord) is made a discord; and by *doubling* a note, is meant the employment of it more than once in the same chord (whether a concord or a discord) in the way of octaves. Thus, in the discord F G B D (in Ex. 1), F is the discordant note (for without it, G B D would be a concord) ; and that note would be *doubled* by playing the chord as F G B D F, as in Ex. 2.

Ex. 1.　　Ex. 2.

---

* Sup. p. 40.

And this, as regards discordant notes in general, is prohibited by the Rule ; an obvious reason for which is, that the discordant sound is by the repetition made too predominant, and consequently too harsh *.

OF CADENCES.

A cadence is a close or termination of a course of melody or harmony. It is like a stop in writing, or a pause in speaking ; and serves, in the same manner as these, to distinguish the ending of any portion or period of the piece, as well as of the whole. Our present concern is with the harmonic cadence. This properly consists of two successive chords ; and it is on the latter of them that the cadence is said to *be*. But it is usually preceded by other chords of an introductory nature, and which form part of the cadence, in a more general sense of the term.

Cadences are of various kinds—Full, Imperfect, or Broken. A Full cadence† is always on the common chord (or, in the minor mode, on the common chord minor) of the tonic ; and it is the only one which is final or conclusive in its effect upon the ear, so as to leave no expectation of any thing to follow.

The Full cadence, in the major mode, consists regularly of an immediate progression from the common chord of the dominant, or of the dominant 7th, to the common chord of the tonic ; the latter chord being in its first position‡, and falling on the first accented note of the bar§. The following is an example of this cadence in its most simple shape :

Ex. 1.

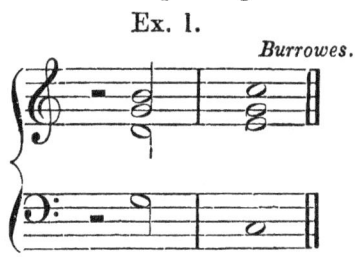

Burrowes.

---

* See Appendix, Note (MM).
† The Full cadence is also called the *Perfect* cadence.
‡ As to the different positions of the common chord, vide sup. p. 32, 52.
§ As to Accent, vide sup. p. 22.

H

But it is often preceded either by the first inversion of the common chord minor upon the 2nd, or by the first inversion of the chord of the 7th upon the 2nd*; of which last progression the following is an example :

Ex. 2.   *Callcott.*

A Full cadence is also further varied by introducing it with the second inversion of the common chord of the tonic, otherwise called the chord of the sixth and fourth; so that the order of the harmonies is, first, that of the tonic (inverted); secondly, that of the dominant; and, lastly, that of the tonic. This species of the Full cadence, which is often denominated *the cadence with* $\frac{6}{4}$, is most usually preceded by the common chord of the subdominant; and the following is an example :

Ex. 3.

In these examples, it will be observed, the common chord of the *dominant* is employed as the penultimate chord. But in the full cadence (whether with or without the $\frac{6}{4}$), the chord of the *dominant* 7*th* is most commonly used instead of the

---

* See Appendix, Note (NN).

common chord of the dominant (with a resolution, of course, in the concluding chord); and the cadence is thereby rendered still more complete in its effect.

The following are examples of Full cadences with the chord of the dominant 7th*.

Ex. 4.

* In most of these examples, an 8th is added to the chord of the dominant 7th. This addition is very usual.

† The Examples bearing this signature are (with some occasional alterations) taken from a Treatise published in 1784, by the Rev. W. Jones, of Nayland, a distinguished amateur.

Ex. 7. *Ibid.*

Ex. 8.

The Full cadence is also sometimes varied by the chord of the diminished 7th, as in the following example:

Ex. 9. *Burrowes.*

and by other methods*.

All these are to be considered as species of the Full cadence in its *regular* form. But the form of the full cadence is sometimes *irregular*. Thus we occasionally find that the concluding chord is preceded, not by the common chord of the dominant, or the chord of the dominant seventh, but by the imperfect common chord, direct or inverted; or by an inversion of the common chord of the dominant, or of the dominant

---

* See Appendix, Note (OO).

seventh; or that it is in the second or third position instead of the first; in all which cases, the cadence deviates in one respect or another from the regular form before described, though it is, nevertheless, a Full cadence, as closing on the common chord of the tonic.

There is also another irregular and somewhat rare form of the Full cadence, in which, though the concluding chord is the common chord of the tonic, yet it is preceded not by the common chord of the dominant, but by that of the sub-dominant; and this is called, in the treatises, the *Plagal* Cadence*; the Full cadence, in its regular form, being distinguished from it as the *Authentic.*

The following is an example of the Plagal cadence—

There is also the Imperfect (or Half) cadence†; which is in the inverse order of the Full cadence; the close being on the common chord of the dominant, preceded by the harmony of the tonic.

Thus: Ex. 1.

Burrowes.

Ex. 2.   *Rogers.*

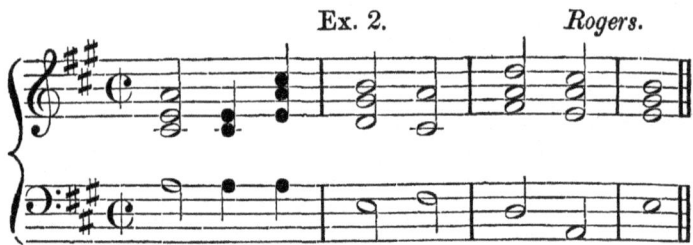

Lastly, there is the Broken (or Interrupted, or False) cadence; which is one that leads to the expectation of the regular Full cadence, but disappoints the ear by varying unexpectedly from that cadence in respect of the progression of the bass of the penultimate chord. For, in the regular Full cadence (as appears by the examples of it before given)*, the bass of the penultimate chord always either falls a fifth or rises a fourth; but, in the cadence now in question, it is made to rise a tone only. Thus:

Ex. 1.   *Callcott.*

Or thus—

Ex. 2.   *Burrowes.*

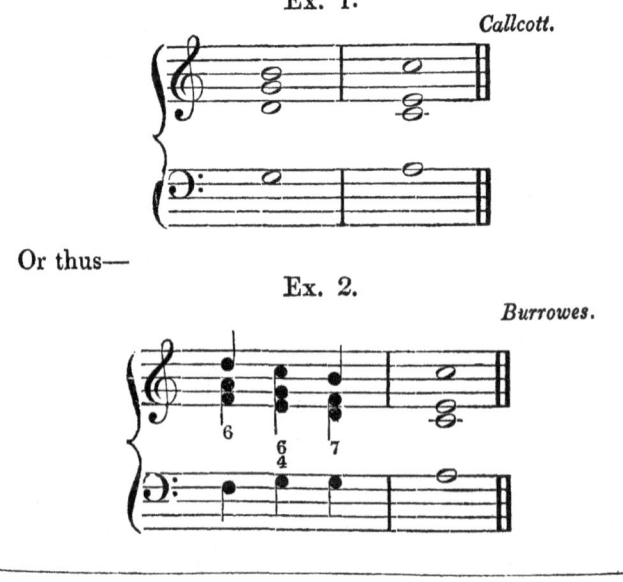

---

* Vide sup. p. 97—100.

We have hitherto considered cadences only in reference to the *major* mode. As to cadences in the *minor* mode, they are made in the same manner as in the major, subject to the following exceptions—that in the Full cadence the concluding chord is the common chord *minor*, that when the discord of the dominant seventh is employed, it has the resolution proper to the minor mode; that is, the subdominat descends a whole tone instead of a semitone*, and that, in theBroken cadence, the bass of the penultimate chord rises a semitone.

The following are examples of the Full cadence in the minor mode:

Ex. 1.

Shield.

Ex. 2.

Ex. 3.

W. Jones.

_____

* Vide sup. p. 95.

Ex. 4.

W. Jones.

The following is an example of the Imperfect cadence in the minor mode.

Ibid.

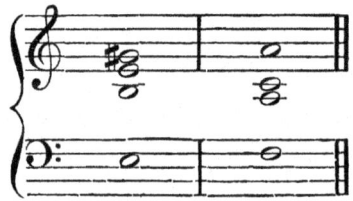

The following, of the Broken cadence in the minor mode.

Callcott.

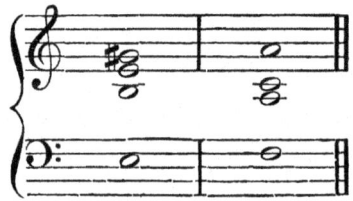

OF MODULATION.

MODULATION is a change from one key to another, in the course of the same piece of melody or harmony. It is to harmonic modulation only, that the following observations relate*.

---

* See Appendix, Note (RR).

Though there is an unbounded variety as to the *particular* manner in which changes of key may be allowably effected, yet they are for the most part founded on one *general* method, viz. that of introducing the chord of the dominant 7th of the intended new key, or one of its inversions; the nature and reason of which expedient will be better understood after a few examples shall have been given.

The most simple and also the most usual modulations from any key are into one of those most nearly related to it, viz. into a key differing only by a single sharp or flat, or having altogether the same notes, though in the alternative mode*.

The following are examples.

Ex. 1.

From C major into G major, by the dominant 7th of the new key.

W. Jones.

Ex. 2.

The same, by the first inversion of the dominant 7th of the new key.

Ibid.

---

* Vide sup. p. 17, and Appendix, Note (SS).

Ex. 3.

The same, by the third inversion of the dominant 7th of the new key.

A. Reicha.

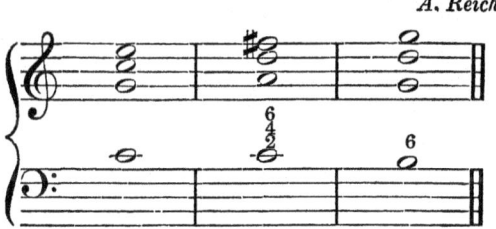

Ex. 4.

From C major into F major, by the dominant 7th of the new key.

W. Jones.

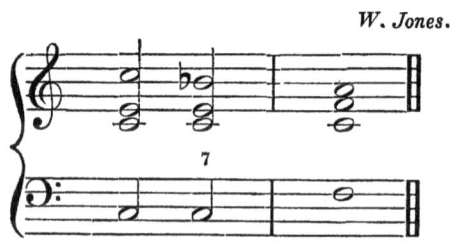

Ex. 5.

The same, by the first inversion of the dominant 7th of the new key.

Ibid.

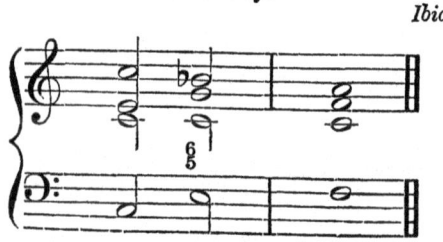

Ex. 6.

The same, by the third inversion of the dominant 7th of the new key.

Ibid.

Ex. 7.

From C major into A minor, by the dominant 7th of the new key.

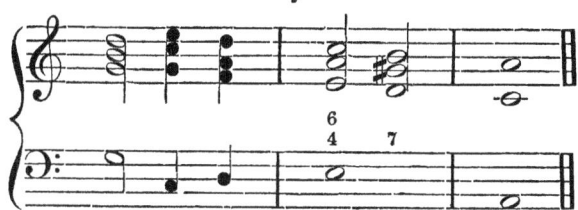

Ex. 8.

The same, by the second inversion of the dominant 7th of the new key.

A. Reicha.

Ex. 9.

From A minor into E minor, by the third inversion of the dominant 7th of the new key.

A. Reicha.

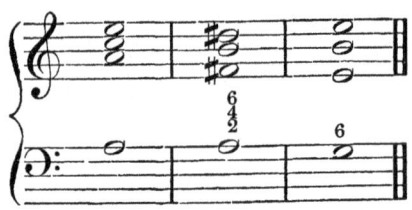

Ex. 10.

From A minor into D minor, by the third inversion of the dominant 7th of the new key.

Ibid.

Ex. 11.

From A minor into F major, by the first inversion of the dominant 7th of the new key.

W. Jones.

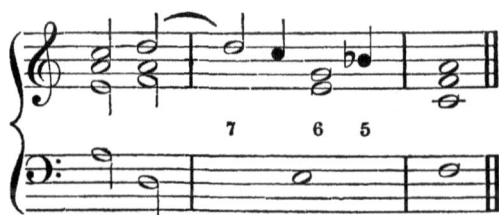

These examples may be analysed as follows :

In the first example, the harmony commences with the common chord of C, the tonic ; in the next chord, the 2nd of the key is taken for a bass, and made to bear the dominant 7th (omitting the 5th and doubling the bass note) in the key of G; and by means of this chord, the new key of G being introduced, the common chord of the tonic of that key accordingly follows in the next bar.

In the second example, the harmony begins with the first inversion of the common chord of the tonic, which is succeeded by the first inversion of the chord of the dominant 7th in the key of G (instead of the direct chord of the dominant 7th, as in the first example), and the key of G being thus introduced, the common chord of the tonic in G follows accordingly in the next bar.

In the third example, the harmony begins with the common chord of the tonic, C, which is followed, on the same bass, by the third inversion of the chord of the dominant 7th of the key of G; and that key being thus established, the first inversion of the common chord of its tonic succeeds.

In the fourth example, the first chord is the common chord of the tonic, C, with the 5th omitted; to which, in the next chord, a minor 7th is added without any other change; and by that mere addition, the harmony is turned into the chord of the dominant 7th of the key of F (with the 5th omitted); by which, the new key of F being introduced, the common chord of the tonic in that key succeeds.

In the fifth example, the same plan of modulation is pursued, except that the first inversion of the chord of the dominant 7th of F, instead of the dominant 7th itself, is employed.

In the sixth example, this is varied by employing the third inversion of the chord of the dominant 7th of F, instead of the second, and by taking as the next chord the first inversion of the common chord of the tonic of the new key, instead of the common chord itself.

In the seventh example, the first chord is the common chord of the dominant, which is followed by the common chord of the tonic, then by the common chord minor of the second, then by the second inversion of the common chord minor of the sixth, then by the chord of the dominant 7th in the key of A, which latter key being thus introduced, the harmony closes on the common chord minor of the tonic, and A in the minor mode is thus established.

In the eighth example, the same key of A is introduced, by the second inversion of the chord of its dominant 7th following immediately upon the common chord of the tonic in the key of C, and the close is in the minor mode as before.

In the ninth example, the harmony begins on the common chord minor of the tonic, in the key of A minor, and the change to the key of E minor is effected by the third inversion of the chord of the dominant 7th of the key of E, which is followed by the first inversion of the common chord minor of its tonic.

In the tenth example, the first chord is the common chord minor of the tonic in A minor, and the change to the key of D minor is effected by the third inversion of the chord of the dominant 7th of the key of D, which is followed by the first inversion of the common chord minor of its tonic.

In the eleventh example, the harmony begins on the common chord minor of the tonic in the key of A minor, which is followed by the common chord of the subdominant, then by the chord of the dominant 7th, suspending the first inversion of the chord of the dominant 7th of the key of F, which immediately follows. The new key of F being thus introduced, the harmony proceeds to the common chord major of the tonic of that key in the next bar, and the key of F major is thus established.

Such, in a practical view, is the nature of modulation in its simplest form ; but, in order to apprehend correctly its *principle* (a subject on which the treatises are generally defective), it will be necessary, in the first place, to take a distinct view of its *object*, which, as already stated, is to change the key. Now this, if closely considered, involves the following ideas : first, an original key ; secondly, a departure from that key ; and thirdly, a substitution of some other.

The first of these ideas leads us to the consideration of the manner in which any key is capable, in the first instance, of being established. And here we must refer to the account, given in the first chapter, of the formation of the successive keys*, from which it may be collected (and it will be further apparent on examination of the Synopsis of keys, sup. p. 15) that no two notes constituting the 4th and 7th, otherwise called the subdominant and leading note, in any key in the major mode, are both to be found in any other key in that mode; and that this is not true of any other two notes in such given key. Thus C, the subdominant, and F♯, the leading note of the key of G, are not both to be found in the scale of any other key than G ; but A and B, which are both in G, are also both in C : C and D, which are both in G, are also both in C ; and F♯ and B, which are both in G, are also both in D; and so of any other two notes, except only C and F♯. These notes then, viz. the subdominant and leading note, jointly considered, are, in each key in the major mode, the *characteristic notes ;* so that the occurrence of both sufficiently establishes that key, to the exclusion of any other ; and, é converso, without the concurrence of both, its establishment cannot be effected.

As to the *minor* mode, if the descending series of its scale only be considered, it will be found that the 3rd and 7th notes descending (which answer to the subdominant and leading note in the major) are, like them, also *characteristic ;* for no two notes, constituting the 3rd and 7th descending in any given key in the minor mode, can both occur in the descending series of any other key in that mode.† But it is impracticable, nevertheless, to obtain, by the employment of the 3rd and 7th descending, so decisive an effect to determine the new key, as by the leading note and subdominant in the case of the major mode, owing to the necessity of taking into account the ascending as well as the descending series, which in this mode, as we have seen‡, are different. Thus, if we are in the key of C minor, and desirous to go into the key of G minor, the employment of the 3rd

---

with the 7th descending in G minor, viz. E♭ and A, will not in itself decisively establish that key ; for E♭ and A are both also found (if we take the ascending series into account) in other keys, viz. C and B♭ minor*.   There are other pairs of notes, however, which, even if the ascending as well as descending series are taken into account, are *charácteristic* in the minor mode ; and these are the leading note and 3rd descending, or the leading note and 6th descending.   Thus, if we are in the key of C minor, and desirous to go into G in the same mode, the employment of F ♯, the leading note in G, with E♭, the 3rd descending, will decisively establish the key of G minor, to the exclusion of all others in that mode ; for no other key in that mode comprises both F♯ and E♭ †. And the employment of F ♯ with B♭, the 6th descending, will be equally decisive for a similar reason, no other key in the mode comprising both F ♯ and B♭.

We see, then, how the object first above pointed out, viz. the establishment of an original key, is accomplished.   As to the result secondly mentioned, viz. a *departure* from that key, it may obviously be effected by the mere introduction of some foreign note or notes.   Thus, if we are playing in the key of C major, the introduction of F ♯ or G ♯ will be a departure ; for neither of these notes are to be found in the original key ; and therefore to that extent the original key is relinquished.

The object, however, which was mentioned in the third place, viz. *the substitution of a new key*, requires something more than a mere departure from the original one.   It requires the employment of the two notes constituting the subdominant and leading note of the new key proposed‡ ; for, without the concurrence of these (as we have seen), no key whatever can be established ; and the concurrence of these is not necessarily involved in the introduction of a foreign note, or even of several foreign notes.   Thus, if it be proposed to change the key from C major to G major, this can be ac-

---

* Vide the Synopsis, sup. p. 16.

† See Appendix Note (JJ).

‡ Vide the Synopsis, sup, p. 16.

complished only by the employment of C and F♯ (the subdominant and leading note of G major), and therefore the
introduction of F♯, though a foreign note, will not decisively
change the key, unless C also be employed. But, on the other
hand, supposing C to be played *contemporaneously* with F♯,
by the introduction of some chord which comprises both, or
supposing C to be played *previously* or *subsequently* (no other
foreign note incompatible with the key of G being allowed to
intervene), the key of C will, on any of these suppositions,
be effectually changed into G.

These things being understood, it is easy to explain (and
first as regards the major mode) upon what principle it is that
the chord of the dominant 7th, or one of its inversions, becomes an instrument of modulation. For it is another result
of the constitution of the scale, that the chord of the dominant
7th of any key (and therefore, of course, any of its inversions)
always comprises the subdominant and leading note of that
key; so that, to establish a decisive change from one key
in the major mode, to another in the same mode, we have
only to employ the dominant 7th of that key, or one of
its inversions ; that necessarily involving the employment of
the subdominant and leading note of such key, which in this
mode are (as we have seen) the *characteristic* notes. Thus D,
F♯, A, C, the dominant 7th of the key of G, comprises its subdominant, C, and its leading note, F♯. Supposing, therefore,
the passage to lie in the key of C major, we can change at
once to the key of G major, by sounding the single chord
D, F♯, A, C. Thus, in the first example*, when the original
key is supposed to be C major, there is a change into G major
by means of the single chord, D, D, F♯, C (equivalent, for
this purpose, to D, F♯, A, C). For the F♯ and the C prove
the key, if the mode be major, to be G. And that it *is* major,
results from the consideration that the passage begun in C
major, and that nothing has occurred to change the mode.

---

* Vide sup. p. 105.

I

Upon the same reason, indeed, it will follow that any *other* chord, comprising both the subdominant and the leading note of the new key proposed, will be equally effectual for modulation into that key; and this is, in fact, the case with respect to the *imperfect common chord*, or its inversions; as may be illustrated by the following example:

where the change from the key of C to that of G is effected by A, F♯, C, the first inversion of F♯, A, C, the imperfect common chord in the key of G. And the case is the same with respect to *the chord of the 7th upon the leading note**, or its inversions; this differing from the imperfect common chord only by the addition of a 7th, and, like it, comprising the leading note and the subdominant of the key proposed. It is true, indeed, that the dominant 7th (direct or inverted) is in much more frequent use in modulation than either of the chords just mentioned; but that is only because, from its having the dominant for its fundamental bass, it is of much superior harmonical importance, particularly for the purpose of a full cadence, and because it is in itself a chord more pleasing than the others to the ear.

It is not within the compass of any *single* chord, however, that the characteristic notes are always introduced. They are often introduced successively, by means of *several* chords, and such a modulation differs from the former in point of effect, by impressing the change of key more gradually upon the ear. A modulation by successive chords is exhibited in the following example:

---

* For this chord, see the Table of Sevenths, seventh column, sup. p. 38.

W. Jones.

where, in changing from C major to F major, the subdominant of the new key, viz. B♭, is introduced in the first bar, and the leading note, E, in the second bar ; and therefore the modulation is effected gradually by several chords, and not at once, as in the fourth, fifth, and sixth examples*, where the change is also from C to F, but by the single chord of the dominant 7th, direct or inverted.

These explanations, it will be observed, relate to the *major* mode only. But, in the *minor* mode, it is usual to modulate through the medium of the same principal agent as in the major, viz. the chord of the dominant 7th, direct or inverted ; its harmonical importance and pleasing character recommending it equally for the purpose in either mode. It is true that in the minor mode the chord of the dominant 7th will not in itself absolutely determine the new key, as it does not contain those which we have shewn to be, in this mode, the characteristic notes, viz. the leading note with the 3rd or 6th descending; but the leading note without either of them. Yet, if it be followed by a chord which contains the descending 3rd or 6th of the new key, that key is gradually established by the chord of the dominant 7th and following chord taken together ; because these chords so taken comprise the characteristic notes required. Thus, in the 9th example†, in the 2nd bar, the chord A, F♯, B, D♯, the 3rd inversion of B, D♯, F♯, A, the chord of the dominant 7th of the new key intended, viz. E minor, contains D♯, the leading note of the new key ; and the chord G, E, B, E, in the next bar, contains G, the 6th descending

---

* Vide sup. pp. 106, 107.   † Vide sup. p. 108.

of E minor; and, therefore, on the whole, the key of E minor is decisively established.

In the minor mode, however, it is practicable, as in the major, to modulate by a *single* chord; and the kind of chord applicable to that service is an extraneous discord. For it is the property of certain chords of this class to comprise the leading note and the 3rd or 6th descending of some particular key in the minor mode; and when such a chord is used, it is in itself sufficient to establish that key. The following is an example of the employment of *the chord of the diminished 7th* for the purpose :

Burrowes.

where the modulation is from the key of C major into the key of A minor; which is effected by the chord of the diminished 7th on the leading note of A minor (being the 2nd chord in the example); for this chord contains both the leading note and the 3rd descending of the proposed key, viz. G♯ and F.

For the easier illustration of the principle of modulation, we have hitherto considered it only in its simplest form; viz. that of modulation into one of the class of keys that are most nearly related to the original one, viz. those differing from it only by a single sharp or flat, or having altogether the same notes, though in the alternative mode; and it is to be observed, that, in this case, the modulation is usually *immediate ;* by which we mean that it is effected without the previous interposition of any other key (as in all the examples above given). But modulation may also take place into a remote key; viz. a key differing from the original one in two or more of its notes. And such a modulation is usually *intermediate ;* that is, effected by the previous interposition of some other key or keys, of such nature and so arranged, that each key throughout the entire series shall be nearly related to the last preceding one.

This may be illustrated by the following example:

(Messiah) Handel.

Get thee up into the high moun- - - -

- - - - - - - - tain.

in which there is a modulation from D major to E major (a remote key), and the intermediate key of A major (nearly related to both) is introduced in the 2nd bar by the chord B, G♯, D, containing the characteristic notes of A major.

An *immediate* modulation into a remote key is also permissible ; though it is much less usual than an intermediate one, and requires more management—in order to prevent the abruptness with which a transition from one key to another, differing by two notes or more, is apt to be attended. Its effect is, in general, best, when it is so managed as to prepare the ear for the change, by introducing some of the new notes of the new key, prior to its complete establishment; as in the following example:

A. Reicha.

where the modulation is *immediate* from C major (which the example supposes to be sufficiently established in the first instance) to B♭ major (a remote key). But prior to the establishment of the latter key (which takes place in the 3rd chord by the introduction of A and E♭, its characteristic notes), one of the new notes belonging to it, viz. E♭, is employed, and serves to prepare the ear for the new key.

The necessity for this sort of preparation of course increases in proportion to the degree of the remoteness of the key into which the modulation takes place. A modulation that is both immediate and unprepared (or, as it may be shortly termed, a *sudden* modulation) into a very remote key, is in general so harsh in its effect as to be inadmissible. There are methods, however, by which the ear may be reconciled to it, and among them is the method of *Enharmonic Transition.*

This is one of the most curious subjects involved in the science of harmony. To explain it, we must premise that any two notes differing from each other only by an enharmonic interval*, are, so far as the impression on the ear is concerned, nearly the same note; and that in keyed instru-

---

* Vide sup. p. 27.

ments they are actually the same, being sounded by the same finger note. Thus G♯ and A♭, which differ only by an enharmonic interval, are nearly identical to the ear, and on a keyed instrument (for example, a pianoforte) absolutely identical; the same finger note expressing both one and the other. Now the method of enharmonic transition, in its ordinary form, consists in placing the chord intended to be used as the instrument of modulation in immediate succession to another chord, which contains a note differing from some note in the chord first mentioned by an enharmonic interval; and by this expedient (supposing the progression to be in other respects properly managed) it is found that a sudden change may be made into a remote key, and even into one exceedingly remote, without giving any offence to the ear, but a surprise only, and a surprise of an agreeable kind. The following is an example :

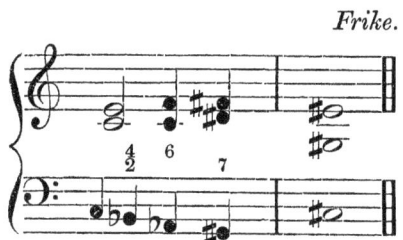

Frike.

where the chord of the dominant 7th in C♯, the chord intended as the instrument of modulation, is placed in immediate succession to the chord of A♭, C, F, belonging to the key of F minor, and containing a note (A♭) which differs from the G♯ in the chord of the dominant 7th only by an enharmonic interval: and we are thus suddenly transported (with a strange but not unpleasing effect) from the key of F minor to the very remote key of C♯ major.

The chord used for the modulation in this example, viz. the chord of the dominant 7th, is one of the *Discords of the Scale ;* but a modulation by an enharmonic transition more frequently takes place through an *Extraneous Discord ;* viz.

either the chord of the diminished 7th, or the chord of the extreme sharp 6th. The following is an example of an enharmonic transition through the latter chord, in a modulation from D♭ into G major:

A. Reicha.

To modulations of every kind the rule particularly applies, which has been already laid down with respect to harmonic progressions in general; viz. that a *liaison* must be maintained between the successive chords*; for this is of great effect in smoothing the passage from one key to the other. Indeed, this is so much the case, that the efficacy of an enharmonic transition to reconcile the ear to a sudden modulation into a remote key, seems to reside altogether in the power that it possesses of creating a *liaison* which would not otherwise exist. For introducing (as it does) a note which differs from some note in the chord last preceding only by an enharmonic interval, and which on a keyed instrument is exactly the same note to the ear, it produces in effect the *liaison* of a note in common with that chord.

Thus, in the example, sup. p. 119, the note G♯, in the chord G♯, D♯, F♯, being substantially the same as the note A♭, in the preceding chord, A♭, C, F, there is between these chords the *liaison* of a note in common, in addition to the imperfect or semitonic *liaison* (as it may be called†), which otherwise exists between them, in respect of the notes F and F♯. And so, in the last example, the note C♯, in the chord E♭, G, B♭, C♯, being substantially the same as the note

---

* Vide sup. p. 92   † Vide sup. p. 93.

D♭, in the preceding chord E, G, B♭, D♭, there is between these chords a *liaison* in respect of this note, in addition to that otherwise existing between them in respect of the notes G and B♭, and in addition to the semitonic *liaison* in respect of the notes E and E♭. And, in both examples, this close connection of the *chords* evidently counteracts the abruptness of the change between the *keys*.

Before we conclude this chapter, it will be necessary to notice two species of progression, which, though not properly modulations, are at the same time of a similar nature.

The first is a transition from one mode to another, the tonic remaining unaltered; as from C major to C minor, or vice versâ. In such a case, as there is no change of *key*, but only of *mode*, there is no modulation, in the strict meaning of that term; but the effect is very similar to that of a modulation; and the manner in which it is effected, the same; viz. by introducing the two characteristic notes which belong to the new key. The following is a transition of this kind, from C major into C minor.

(*The Surprise*) *Haydn.*

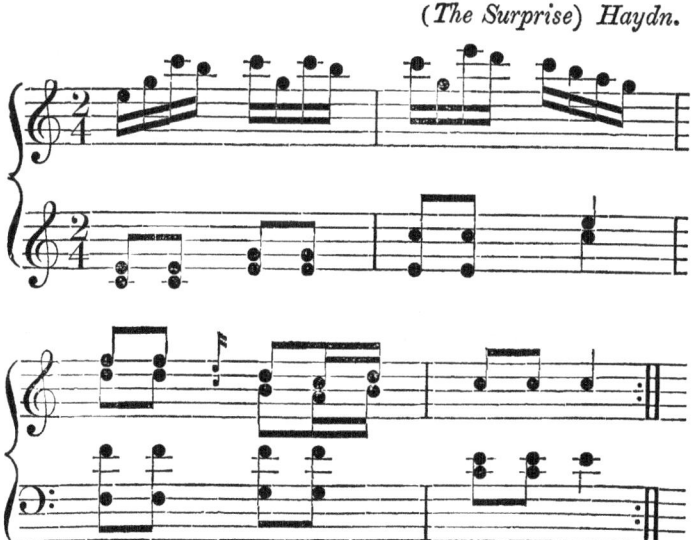

The remaining progression is that which is commonly described as a *chromatic passage*, which is a succession of chromatic semitones, usually intermixed with semitones of the diatonic kind*.

A chromatic passage sometimes involves a change of key; for that is one of the methods by which a chromatic semitone may be produced :—as in the following example :

Callcott.

---

* As to chromatic and diatonic semitones, vide sup. p. 28.

where the intervals from D♯ in the 1st chord to D♮ in the 2nd, from G♯ in the 2nd to G♮ in the third, from C♯ in the 3rd to C♮ in the 4th, and from F♯ in the 4th to F♮ in the 5th, are all chromatic semitones (intermixed with the diatonic semitones from D♮ in the 2nd bar to C♯ in the 3rd, from G♮ in the 3rd to F♯ in the 4th, and from C♮ in the 4th to B in the 5th), and are produced by modulating in succession through different keys; viz. from E to A, from A to D, from D to G, and from G to C, by the employment, at each step, of the chord of the dominant 7th of the next key.

But a chromatic passage may also take place without any change of key; viz. by the interchange, in the course of the same *minor* key, between the augmented and unaugmented 6th and 7th notes of the ascending series. The following is an example of a chromatic passage produced in this manner:

W. Jones.

where the intervals from G♯ to G♮ in the 2nd bar, and from F♯ to F♮ in the 3rd, are chromatic semitones, and are produced by employing alternatively G♯ and G♮, the augmented and unaugmented 7th, and F♯ and F♮, the augmented and unaugmented 6th, in the course of the single key of A minor.

Chromatic passages are of very frequent occurrence in music; but are never so appropriately and powerfully employed as in the expression of feelings of a mournful or dejected kind; an illustration of which may here be given from the works of a great master.

CORO.   Pergolesi.

CHAPTER IV.*

OF CHORDS, CONSIDERED INDIVIDUALLY.

I. *The Common Chords, Major and Minor.*
(Table, sup. p. 53, Nos. 1, 4.)

THOUGH, in the present work (as in some others), it has
been thought desirable, in point of arrangement, to class the
chord of minor 3rd, minor 5th and 8th, upon the leading
note, as one of the common chords of the key, and consequently to assign to it the appellation of imperfect common
chord; yet in its nature it differs very materially from the
common chords, major and minor, being a discord, while
these are concords. On the present occasion, therefore, it
will be convenient to treat of the imperfect common chord
separately, and in a subsequent place, confining our present
attention to the common chords, major and minor.

We have already had occasion to state the intervals of
which the common chord, whether major or minor, consists†,
and to refer to that remarkable property of the common chord
major by which its several notes are referable to one common

---

* This chapter should be read in connection with the Table of Chords,
sup. p. 53.

   † Vide sup. pp. 31, 33, and Table, sup. p. 53.

generator or fundamental bass*; and we have shewn how the idea of a fundamental bass has become extended to the common chord minor also, and to some others†. We have moreover explained the different *positions* and *inversions* to which common chords, both major and minor, are subject‡; and have shewn upon what notes of the key these several chords may respectively be taken§.

We may now farther remark, in the first place, that the common chord major is characterized by a certain unity, fulness, and boldness of sound; and that the common chord minor, though very similar to it in general effect, is somewhat less distinguished by these several qualities. The character of these chords may be illustrated by the following examples, from the two greatest masters of harmony:

Ex. 1.

(Messiah) Handel.

And his name shall be called

Wonderful! Counsellor!

---

* Sup. pp. 32, 33. † Sup. p. 34. ‡ Sup. pp. 32, 34.
§ Sup. p. 35.

The mighty GOD, the e - ver - last - ing FATHER, the PRINCE of PEACE.

Ex. 2.

(*Creation*) *Haydn.*

And God said: Let there be

And God said: Let there be

And God said: Let there be

And God said: Let there be

The first of which illustrious passages consists almost wholly of the common chord major; the second introduces, under common chords minor*, the drawing of the nascent light; and then pours its flood upon us in a burst of common chords major†.

---

* In the 2nd, 3rd, and 4th bars.   † In the 6th and 7th bars.

The positions of the common chord (whether major or minor) are 3 in number, according as the 8th, 3rd, or 5th is uppermost; as will be fully explained by the following example, which embraces also the imperfect common chord.

The most useful of these positions is that which is commonly called the 1st, and which is also put first in the example; viz. that in which the 8th is uppermost (as C, E, G, C̄), where the bass note is heard both at the bottom and top of the chord. For in that position the common chord may be played as an accompaniment to the key-note*. But in the practice of harmony, frequent recourse is had to *all* the positions; and it is a rule, that when several common chords are taken in succession, their positions should be varied; of which the example affords an illustration.

One of the notes of the common chord may be omitted; and to make a chord of four, in that case, one of the others may be doubled. But it is better, generally, to omit the 8th or the 5th (particularly the 8th), than the 3rd; and where the 8th is omitted, it is better to double the 5th than the 3rd.

---

* See Appendix, Note (TT).

II. *The Inversions of the Common Chord Major and Minor.*

(V. sup. Table, p. 53, Nos. 2, 3, 5, 6.)

The inversions of the common chord (whether major or minor) are two in number*; the first inversion being called the chord of the 6th (figured $\begin{smallmatrix}8\\6\\3\end{smallmatrix}$, $\begin{smallmatrix}6\\3\end{smallmatrix}$ or 6), and the second inversion being called the chord of the 6th and 4th, and figured $\begin{smallmatrix}8\\6\\4\end{smallmatrix}$ or $\begin{smallmatrix}6\\4\end{smallmatrix}$. The former is produced by taking for the bass note that note which is the 3rd of the common chord ; the latter, by taking for the bass note that note which is the 5th of the common chord.

We have already seen, in a preceding part of this work (v. sup. p. 33), that any common chord and its inversions have all the same fundamental bass (viz. the bass note of the common chord), though they have all different bass notes. When a common chord occurs, its fundamental bass is at once apparent, being the same with its bass note. But when an inversion occurs, the fundamental bass, being always different from its bass note, is not so immediately discernible. Thus, upon sight of the chord  which we know, by its intervals of major 3rd, 5th and 8th, to be a common chord, we also know at once its fundamental bass to be G; for that is its bass note. But upon sight of the

chord 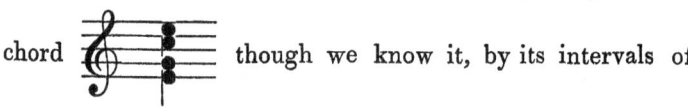 though we know it, by its intervals of

---

minor 3rd, minor 6th and 8th, to be a chord of the 6th, viz. the first inversion of some common chord; we may, nevertheless, pause upon the question, what is its fundamental bass; and it is consequently convenient to be in possession of some rule for its immediate determination. The following is a rule for the purpose; viz. that, in the case of a first inversion, the fundamental bass is a 3rd below the bass note (a *major* 3rd, if the 3rd of the inversion be *minor*, otherwise a *minor* 3rd); and in the case of a second inversion, a 5th below. In the example above given of a first inversion, viz. E, G, C, E, where E, G, the 3rd of the inversion, is minor, the fundamental bass is consequently C; being a 3rd major below the bass note.

We shall now remark on the two inversions severally and in their order.

1st. *The Chord of the 6th.*

This chord, as we have just seen, is produced by inverting the common chord, whether major or minor; besides which, there is another species of 6th (as we shall see hereafter), produced by the inversion of the imperfect common chord. That chord of the 6th, which is produced by the inversion of a common chord major, has for its bass note (in the major mode), the 3rd, 6th, or leading note of the key, and its intervals are a minor 3rd, a minor 6th, and an 8th. That which is produced by the inversion of a common chord minor, has for its bass note (in the major mode) the tonic, subdominant, or dominant, and its intervals are a major 3rd, major 6th, and an 8th.

Among these, the chord of the 6th on the subdominant deserves particular notice, on account of the frequent service that it renders as a chord introductory to a Full Cadence.

An example of its employment in this manner was formerly given under the proper head.*

Of a chord of the 6th, it may be laid down generally, that when it is played full, that is, with all its intervals of 3rd, 6th, and 8th, the 8th note should occupy one of the two middle parts of the chord, and not the uppermost part; and that the effect is best when the 8th note is omitted altogether, and in lieu of it the 3rd or 6th doubled. But both these rules are subject to frequent exception†.

2nd. *The Chord of the* $\frac{6.}{4}$

This chord, as we have seen, is produced, like that of the 6th, by inverting the common chord—whether major or minor : and it may also be produced by inverting the imperfect common chord. Its intervals are generically a sixth and a fourth, though (as in the case of the chord of the 6th) the intervals will vary specifically, according to the bass note on which the chord is taken.

But the chord of the $\frac{6}{4}$ (in the major mode) principally occurs either on the *tonic* or the *dominant* of the key; in which case, it is the inversion of a common chord *major ;* and its intervals in such cases consist of a 4*th* and a *major 6th*.

It is a chord of great use in harmonical progression, as by means of it, we produce that favorite termination of a musical period called the *Cadence with the 6th and 4th* ‡.

As to this chord, and every other in which a 4th occurs between the bass note and one of the upper notes, it is a general rule that the 4th must be *prepared* by the introduction into the previous chord of one of the notes of which such 4th consists §.

---

* Vide sup. p. 99, Example 6. See Note, Appendix (UU).

† See Note, Appendix (VV).　　　‡ Vide sup. p. 98.

§ See Note, Appendix (WW).

Thus, in the following example,

Heck.

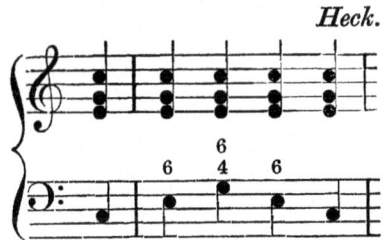

the 4th, G C, in the chord marked $\frac{6}{4}$, formed between the bass G and one of the upper parts, C, is prepared by the occurrence of the C in the preceding chord.

And so, in the following example,

Heck.

the 4th, G C, in the first of the chords marked $\frac{6}{4}$, is prepared by the occurrence of the bass note G in the preceding chord; and the case is the same with regard to the second of the chords marked $\frac{6}{4}$.

But there is an exception to this rule in the case of *Cadences;* the reason for which exception is, that the deviation in this case from the ordinary rule gives greater force and emphasis to the cadence.

Thus, in the following half-cadence,

A. Reicha.

and in the following full cadence,

A. Reicha.

we may observe that the 2nd chord in each case (being a chord of $\frac{6}{4}$) has a 4th, D G, between the bass and an upper part—for which no preparation is made; and yet the progression is good.

It is also to be observed, that the rule has no application to the case of a 4th formed not between the bass and an upper part, but between two upper parts, as in the case of

the chord of the 6th, E C G C  where there is

a 4th, G C, between the 1st and 2nd parts, but none between the bass and any upper part.

III. *Chord of the* $\frac{8}{5}$ *or* $\frac{5}{4}$.

(V. sup. Table, p. 55, No. 7.)

In the ancient music, this chord was more frequently used than it is at present. It is now chiefly employed as preparatory to a cadence; in which case it is placed upon the dominant.

Grave.   (*Te Deum*) *Handel.*

Let me   nev - er   be con - found - ed.

Other examples of this chord have already been given under the head of Cadences.

IV. *Chord of the* $\frac{5}{2}$.

(V. sup. Table, p. 55, No. 8.)

To make a chord of four notes, the 2nd or 5th of this chord may be doubled. The following is an example of the $\frac{5}{2}$ when placed on the tonic.

Corfe.

V. *Chord of the* $\begin{smallmatrix}5\\4\\2\end{smallmatrix}$.

(V. sup. Table, p. 55, No. 9.)

The following is an example of this chord when placed on the tonic.

Heck.

VI. *The Imperfect Common Chord, with its Inversions.*

(V. sup. Table, p. 55, No. 10.)

This chord is called by some writers the *diminished triad.* Its bass note is the leading note of the key in the major mode, and the second ascending in the minor ; and its intervals, a minor 3rd, a minor 5th, and an 8th*. And as the minor 5th is a discordant interval, the whole chord must consequently be ranked as a discord†.

Like the common chord, it has two inversions, the first produced by taking for the bass note the 3rd, and the second by taking for the bass note the 5th, of the direct chord. The first (figured $6, \begin{smallmatrix}8\\3\end{smallmatrix} \begin{smallmatrix}6\\3\end{smallmatrix}$) consists of a minor 3rd, major 6th, and 8th.

The second, (figured $\begin{smallmatrix}8\\6\\4\end{smallmatrix}$ or $\begin{smallmatrix}6\\4\end{smallmatrix}$) consists of a tritone, major 6th, and 8th.

This chord as occurring in the major mode, will be found in each of its different positions in the third bar of the Example, sup. p. 130. And its several inversions as occurring in the minor mode, are exhibited in the third and sixth bars of the following Example.

---

* V. sup. p. 33.　　　　† V. sup. p. 39.

138

Shield.

The fundamental bass of the imperfect common chord, and of its several inversions, is the leading note of the key* in the major mode, and the second ascending, in the minor.

This chord, as noticed in a former place, has the property (in the major mode) of comprising the *characteristic* notes of the key; and by the mere use of it, a modulation is consequently capable of being effected; though the chord ordinarily used for that purpose is another having the same property; viz. the chord of the dominant 7th†.

VII. *The Chord of the Dominant 7th.*

(V. sup. Table, p. 57, No. 11.)

This chord has for its intervals a major 3rd, 5th, and minor 7th, with the occasional addition of an eighth.

---

* Vide sup. p. 36.   † Vide sup. pp. 105, 113, 114.

It may be considered, therefore, as the common chord major, with a minor 7th (which is a discordant interval) added.

It is called the chord of the *dominant* seventh, because it is only on the *dominant* of the key as a bass, that such a chord can (in the major mode) be taken; as will be at once apparent from the Table, sup. p. 38, where it will be observed that, in the key of C major, chosen for the example, it is only the *dominant*, 7th, G B D F, that comprises the exact combination of major 3rd, 5th, and minor 7th.

Various examples of the chord of the dominant 7th have been given, under the head of Cadences*; to which may now be added the following:

(*Fidelio*) *Beethoven.*

* Vide sup. p. 99.

In the Table of chords, as well as in the above example, the different *positions* of this chord are also exhibited.

The manner of its resolution has been explained in a former place*.

The fundamental bass of the chord of the dominant 7th is the dominant of the key.

The 7th is the characteristic note of this chord (as of every chord of the 7th), and therefore cannot be omitted; nor (as it is a dissonant note) can it in general be doubled.

It is also to be observed, as to the progression of this chord, that it is better to *descend* upon the discordant note, as in the 1st of the following examples, than to *ascend* to it, as in the 2nd.

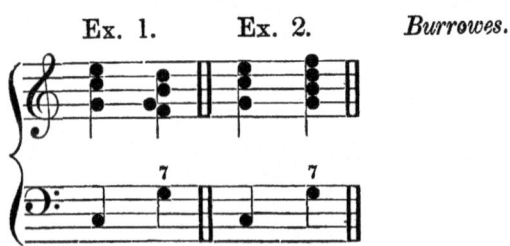

Ex. 1.   Ex. 2.   *Burrowes.*

This chord is of a very remarkable and important character; for it possesses (as has been already shewn under

---

* V. sup. p. 94, 95.

141

other heads) the following properties. 1st. It is essential
to the absolute completeness of a full cadence*. 2nd. It
decides (whether taken in its direct or in its inverted form)
the key of the passage in which it occurs†. And 3rd. It
constitutes (in its direct or inverted form) the principal medium through which a modulation is effected‡. We may
add, that it is, of all the discords (intrinsically considered),
one of the most agreeable to the ear, having no rival in this
respect, except, perhaps, the chord of the 9th, taken on the
dominant as a bass.

VIII. *Inversions of the Chord of the Dominant 7th.*

(V. sup. Table, p. 57, Nos. 12, 13, 14.)

The chord of the dominant 7th has three inversions; the
first, figured $\frac{6}{5}$, consists of a minor 6th, minor 5th, and minor
3rd; the second, figured $\frac{6}{3}$ or $\frac{4}{3}$, consists of a major 6th, 4th
and minor 3rd; the third, figured $\frac{6}{4}$, $\frac{4}{2}$, or 2, consists of a
major 6th, tritone, and 2nd. The first is produced by taking
for the bass note that note which is the 3rd of the direct chord;
the second, by taking for the bass note that note which is the
5th of the direct chord; the third, by taking for the bass note
that note which is the 7th of the direct chord. The fundamental bass of the direct chord being the dominant, the
same also is the fundamental bass of its three inversions.

We will remark on the inversions severally and in their
order.

---

* V. sup. p. 97, 99.　　† V. sup. p. 113.　　‡ Ibid.

1st. *The Chord of the* $\frac{6}{5}$.

This chord, viz. the first inversion of the chord of the dominant 7th, has always for its bass (in the major mode) the leading note of the key.

This chord is to be distinguished from other chords of the same denomination and figuring, but placed on other notes of the key, and whose intervals therefore, though generically the same (as consisting of 6th, 5th, and 3rd), are specifically different; among which may be particularly noticed that chord of $\frac{6}{5}$ which (in the major mode) has the *subdominant* for its bass, and whose intervals are a major 6th, 5th, and major 3rd. This last-mentioned chord is not an inversion of the dominant 7th, but the first inversion of one of the other 7ths of the key; viz. the 7th on the 2nd of the key*; and it is a chord of importance, being frequently employed as introductory to a Full Cadence, of which an example has been given in the proper place†.

2nd. *The Chord of the* $\frac{6}{4}$ $\frac{4}{3}$.

This chord, viz. the second inversion of the chord of the dominant 7th, has always for its bass note (in the major mode) the 2nd of the key. But chords of the same denomination and figuring, with intervals somewhat different, may occur on other notes of the key. Thus a chord of $\frac{6}{4}$ $\frac{4}{3}$ may occur on the *subdominant* (in the major mode), being the second inversion of the chord of the 7th on the leading note‡; and its intervals will then be a major 6th, a tritone, and a major 3rd.

---

* V. sup. Table, p. 57, No. 16.

† V. sup. p. 98, Ex. 2. See Note, Appendix (**XX**).

‡ For this chord, see the Table of Sevenths, seventh col. sup. p. 38.

As, in a chord of $\frac{6}{4}$, a 4th occurs, and its place is between
the bass note and one of the upper parts,—it is subject
to the rule already noticed*, in the case of the chords of
the $\frac{6}{4}$; viz. that the 4th must be *prepared* by the occurrence
in the preceding chord of one of the notes of which the 4th
consists.

The following examples will serve to illustrate both
the $\frac{6}{5}$ and the $\frac{6}{4}$; the former, both as occurring on the leading note and on the subdominant; the latter, both as occur-
ring on the 2nd and the subdominant.

Ex. 1.

---

* V. sup. p. 133.

144

Ex. 2.

(*Fidelio*) *Beethoven.*

3rd. *The Chord of the* $\frac{6}{4}$.
$\quad 2$

(V. sup. p. 42, Table, p. 57, No. 14.)

This chord, viz. the third inversion of the dominant 7th has always for its bass note the subdominant of the key.

It is to be distinguished from other chords which have the same denomination and figuring, but which, being placed on other notes of the key, have intervals specifically different; particularly the $\frac{6}{4}$ on the tonic, which consists of a major 6th $\quad 2$ a 4th, and a 2nd. This last is not an inversion of the dominant 7th, but the third inversion of one of the other 7ths of the key; viz. the 7th upon the 2nd of the key*.

The chord of the $\frac{6}{4}$ on the subdominant (which is a chord $\quad 2$ of great effect) will be illustrated by the following examples:

Ex. 1.

* V. sup. Table, p. 57, No. 18.

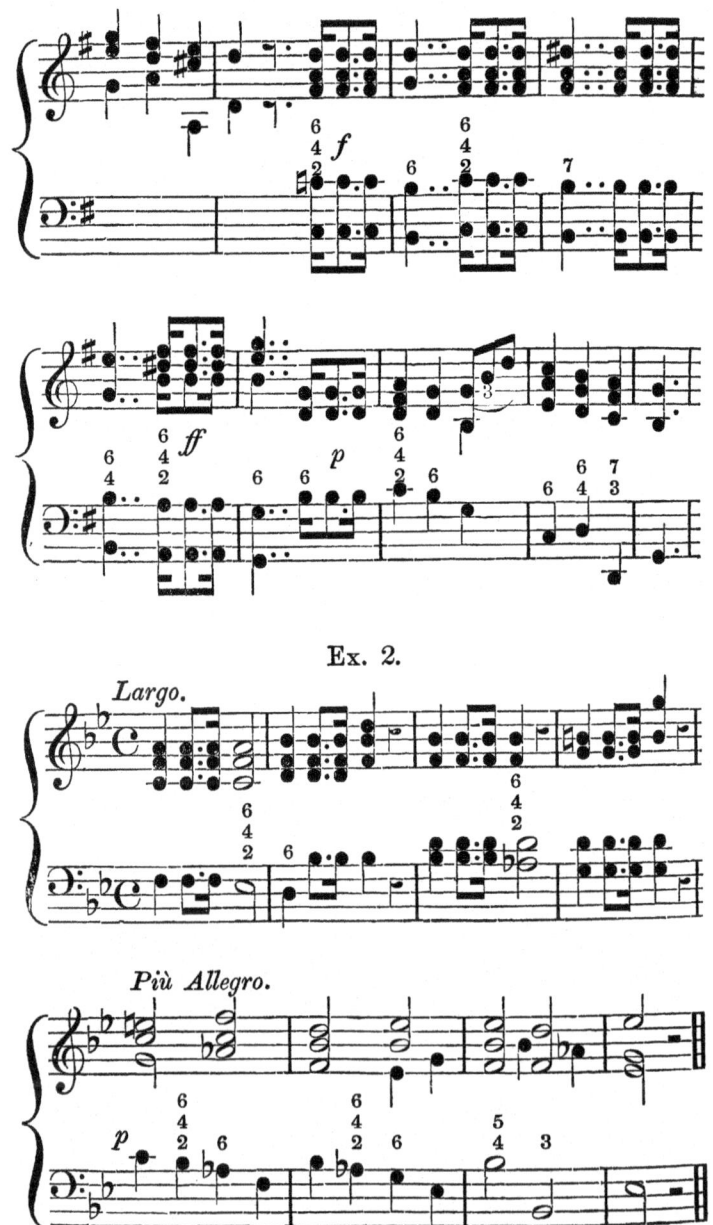

Ex. 2.

Largo.

Più Allegro.

Ex. 3. *Handel.*

The voice of him that cri-eth in the wil-der-ness, "Pre-pare ye the way of the Lord, make straight in the de-sert a high-way for our God."

* The first of these chords of 4 is on the subdominant in the key of

$$\frac{6}{4}$$
$$2$$

E major; the second, on the subdominant in the key of C♯ minor, into which there is a modulation.

Of the $\begin{smallmatrix}6\\4\\2\end{smallmatrix}$ on the tonic, the following is an example.

Ex. 4.

IX. *Other Chords of the 7th.*

(V. sup. Table, p. 57, No. 15.)

Besides the chord of the 7th on the dominant, which is beyond comparison the most important, there is on each of the other notes of the key a chord of the 7th figured in the same way as that of the dominant 7th, and consisting of the same intervals, generically considered, viz. a 3rd, a 5th, and a 7th; but the quality of their intervals, specifically considered—that is, considered as major or minor,—is different, as will appear on examination of the Table, sup. p. 38. Such of them (in the major mode) as occur on the tonic and subdominant respectively, for example C E G B and F A C E, consisting of a major 3rd, a 5th, and major 7th; such as occur on the 2nd, 3rd, and 6th respectively, for example D F A C, E G B D, A C E G, consisting of a minor 3rd, 5th, and minor 7th; and that which occurs on the leading note, for example B D F A, consisting of a minor 3rd, minor 5th, and minor 7th. Of these, it has been thought sufficient to select for insertion in the General Table* the chord of the 7th upon the 2nd of the key.

Each of these chords of the 7th has also its three different inversions produced and figured respectively in the same manner as the inversions of the dominant 7th, and having

---

* V. sup. p. 57, No. 15.

for their fundamental basses the bass notes of the direct chords respectively.

The following example exhibits the chord of the 7th not only on the 2nd, but also on the subdominant, the dominant, and the leading note.

(*Creation*) *Haydn.*

$$\text{X. } \textit{The Chord of the } \begin{smallmatrix}7\\4\\3\end{smallmatrix}.$$

(V. sup. Table, p. 59, No. 19.)

The following is an example of this chord when taken on the tonic; in which case it differs *specifically* from the chord in the Table.

Allegro.

XI. *The Chord of the* $\frac{7}{6}$.
3

(V. sup. Table, p. 59, No. 20.)

This chord (when taken on the dominant) is also illustrated in the last example.

XII. *The Chord of the* $\frac{7^*}{5}$.
4

(V. sup. Table, p. 59, No. 21.)

The following is an example of this chord when taken on the leading note ; in which case it differs *specifically* from the chord in the Table.

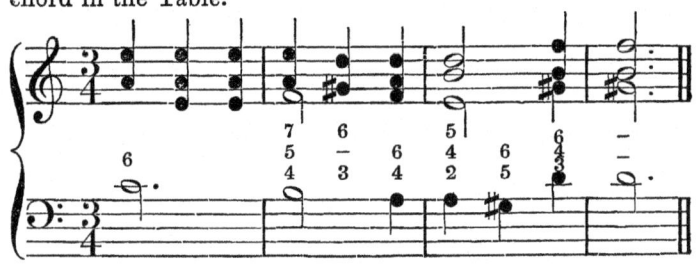

XIII. *The Chord of the* $\frac{7}{4}$.
2

(V. sup. Table, p. 59, No. 22.)

This chord usually occurs on the tonic of a major or minor key, and in passages where the tonic is made a continued or holding note. When so occurring, it may be considered as the chord of the dominant 7th, with the tonic substituted for its proper bass. It resolves into the common chord of the tonic.

An example of the chord $\frac{7}{4}$, occurring in its usual man-
2
ner, will be found sup. p. 139, to which we may add the fol-

---

* The 3rd is sometimes added to this chord, as in the example, sup.
p. 147, *ad finem* ; making a chord of $\begin{smallmatrix}7\\5\\4\\3\end{smallmatrix}$.

151

lowing, where it occurs in a different manner, and is taken on the subdominant and 2nd of a minor key.

Heck.

XIV. *The Chord of the* $\frac{7}{5}$ $\frac{5}{4}$ $\frac{}{2}$.

(V. sup. Table, p. 61, No. 24.)

This chord usually occurs in the same manner as the $\frac{7}{4}$ $\frac{}{2}$, and may be considered as the chord of the dominant 7th (or one of its inversions), with the tonic added as a bass. It resolves into the common chord of the tonic. The following is an example of this chord.

Gunn.

XV. *The Chord of the* $\frac{7}{6}$ $\frac{6}{4}$ $\frac{}{2}$.

(V. sup. Table, p. 61, No. 23.)

This chord also usually occurs in the same manner as the $\frac{7}{4}$ $\frac{}{2}$, and may be considered as the chord of the 7th on the leading note (or one of its inversions), with the tonic added as a bass. It resolves into the common chord of the tonic.

152

The following are examples.

staccato   legato

(*Theodora*) *Handel.*

O worse than death in - deed!

XVI. *The Chord of the 9th.*
(V. sup. Table, p. 61, No. 25.)

The chord of the 9th has generically the intervals of 9th, 5th, and 3rd; but the intervals (as in the case of all other chords) will vary specifically according to the note of the key on which the chord is taken. If on the tonic, or subdominant, or dominant, the intervals are a major 3rd, 5th, and major 9th. If on the 2nd or 6th, a minor 3rd, 5th, and major 9th. If on the 3rd, a minor 3rd, 5th, and minor 9th. If on the leading note, a minor 3rd, minor 5th, and minor 9th.

The 9th on the tonic, 2nd, 3rd, subdominant, and dominant, all occur in the following examples.

Ex. 1.

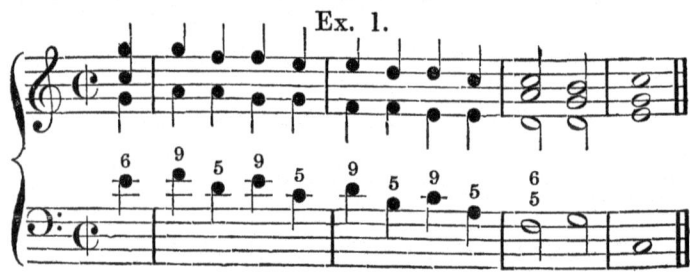

Ex. 2.

Jousse.

Ex. 3.

Jousse.

XVII. *The Chord of the* $\begin{smallmatrix}9\\6\\3\end{smallmatrix}$.

(V. sup. Table, p. 61, No. 26.)

The following are examples of this chord, taken on the 3rd of the key, both in the major and the minor mode.

Heck.

XVIII.   *The Chord of the* $\frac{9}{6}$.

(V. sup. Table, p. 63, No. 27.)

The following are examples of this chord, taken on the 2nd of the key, both in the major and minor modes; in which case it differs *specifically* from the chord in the Table.

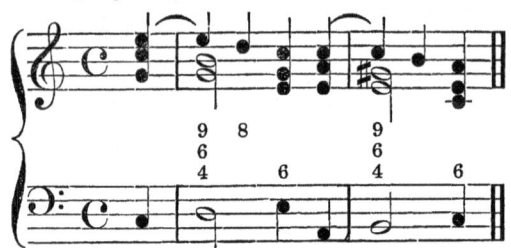

XIX.   *The Chord of the* $\frac{9*}{5}$.

(V. sup. Table, p. 63, No. 28.)

The following are examples of this chord.

---

* The 7th is sometimes added to this chord, making it a chord of $\frac{9}{7}$ $_{5}$ $_{4}$

XX.   *The Chord of the* $\overset{\displaystyle 9^{*}}{\underset{\displaystyle 3}{7.}}$

(V. sup. Table, p. 63, No. 29.)

The following are examples of this chord, taken on the subdominant, 3rd, and 2nd, in the major mode; in which cases respectively the intervals are *specifically* different.

Heck.

XXI.   *The Chord of the* $\overset{\displaystyle 9}{\underset{\displaystyle 4}{7.}}$

(V. sup. Table, p. 63, No. 30.)

The following is an example of this chord, taken on the tonic in the minor mode; in which case it differs *specifically* from the chord in the Table.

Haydn.

---

* The 5th is sometimes added to this chord, making it a chord of $\begin{smallmatrix}9\\7\\5\\3\end{smallmatrix}$

XXII. *The Chord of the Major 7th and Augmented 2nd.*

(V. sup. Table, p. 65, No. 31.)

The following is an example of this chord; which, as well as all those to be subsequently noticed, is of the class of Extraneous Discords.

XXIII. *The Chord of the Minor 6th and Diminished 4th.*

(V. sup. Table, p. 65, No. 32.)

The following is an example of this chord.

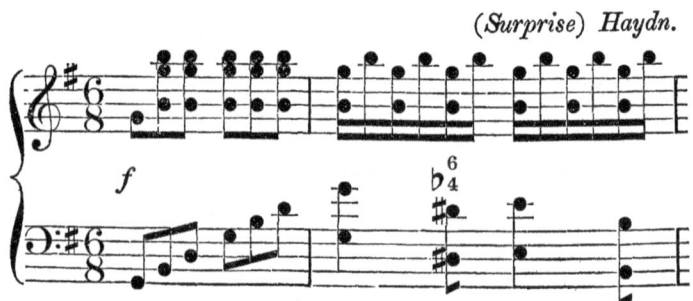

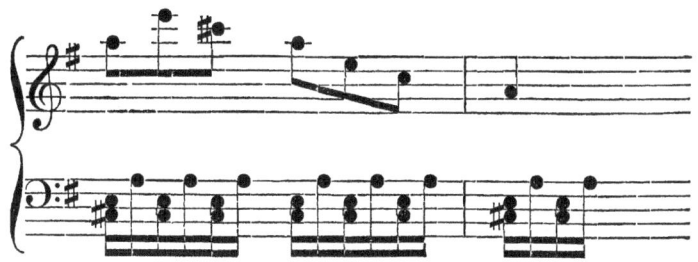

XXIV. *The Chord of the Minor 6th and Major 3rd.*
(V. sup. p. 46, Table, p. 65, No. 33.)

The following is an example of this chord.

Heck.

XXV. *The Common Chord with an Augmented 5th.*
(V. sup. Table, p. 65, No. 34.)

The following are examples of this chord*.

Ex. 1.

---

* See also another example, in the 8th bar of the passage from Haydn, sup. p. 149. In these examples, it will be observed that the note forming the augmented 5th is in a manner *prepared* by the previous introduction of the same note without the augmentation.

Ex. 2.   *Haydn.*

XXVI.   *Chord of the Extreme Sharp 6th**.

(V. sup. p. 46, Table, p. 67, No. 36.)

This chord has always the interval of an augmented major 6th formed between the bass and an upper note ; and

---

* This is otherwise called the chord of *the augmented 6th,* or of *the superfluous 6th.*

when three notes are wanted, it has also the interval of a major 3rd formed between the bass and an upper note. To make a chord of four notes, the upper note of the interval of the 3rd is doubled. Instead of this, however, the bass is sometimes doubled for the same purpose.

This chord, when consisting of three or of four notes, resolves into a chord of $\frac{8}{3}$; the bass note falling a semitone, the note which forms a third with the bass note also falling a semitone, and the note which forms an augmented major sixth with the bass note rising a semitone, as in the example in the Table, sup. p. 66, No. 36, where the chord, F A D ♯, resolves into E G ♯ E, by F's falling a semitone to E, A's falling a semitone to G ♯, and D ♯'s rising a semitone to E. And, supposing the chord to consist of two notes only, it will resolve on the same principle into a simple 8th.

The chord of the extreme sharp 6th is a remarkable chord, as being not only of the class of Extraneous Discords, but one of a more *strictly* extraneous character than belongs in general to chords of that class; for these, as explained in a former place*, are for the most part to be found among the intervals of the *minor* mode, if its *ascending* scale with the accidental augmentations be taken into the account; but the chord of the extreme sharp 6th is foreign to the scale altogether, both in the major and in the minor mode; there being no key, either major or minor, in which its interval of augmented major 6th is to be found†.

The chord of the extreme sharp 6th, though, for the reason just stated, belonging to neither mode, yet is capable of being produced by taking the first inversion of the common chord minor of the subdominant in any minor key, and augmenting the subdominant chromatically. Thus, if we take F A D (which is the first inversion of D F A, the common chord minor of D, the subdominant in A minor), and augment D into D ♯, we have F A D ♯, which is a chord of the extreme

---

* V. sup. p. 48.   † See Note, Appendix (YY).

sharp 6th. The chord so produced is also capable of being resolved upon the dominant of the same key, so as to make a species of Imperfect cadence in that key*. Thus (as we have seen), the resolution of F A D♯ is into E G♯ E, which is a chord on the dominant in A minor. The chord of the extreme sharp 6th is therefore so far related to the minor mode; but it has no such relation to the major†.

The chord of the extreme sharp 6th is also remarkable for its powerful and pungent effect, and appears to be a favorite with all the great harmonists. We select for its illustration the following examples.

Ex. 1.   *Rev. W. Jones.*

Ex. 2.

(*Nozze di Figaro.*) *Mozart.*

---

* The ordinary species of Imperfect cadence (as we have seen) is a cadence on the dominant by a different progression, viz. a progression from the common chord of the tonic. V. sup. pp. 101, 104.

† See Appendix, Note (ZZ).

Ex. 3.   (*Sinf.* 2)   *Haydn.*

M

162

Ex. 4.

(*Esther*) *Handel.*

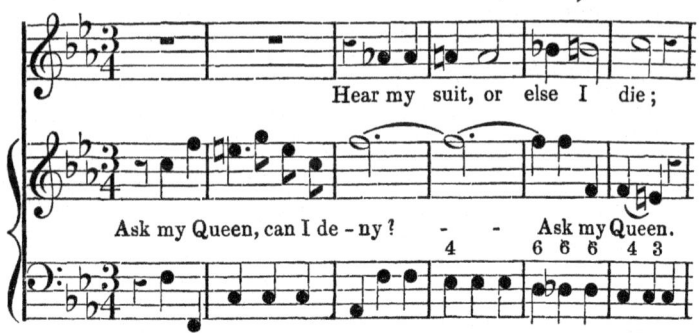

Ex. 5.

(*Creation*) *Haydn.*

God. The wonder of his works.

glo - - - ry of God.

God. The wonder of his works.

God. The wonder of his works.

Ex. 6.

(*Miserere*) *Francesco Ciampi.*

Ec - ce enim in iniqui-

Ec - ce enim

м 2

ta - - - - - - ti - bus con-

in iniqui - ta - - - tibus con-

cep - tus sum!

cep - tus sum!

XXVII. *The Chord of the Extreme Sharp 6th with a 5th.*

(V. sup. Table, p. 67, No. 37.)

This chord also is strictly extraneous ; and differs from the chord of the extreme sharp 6th only by adding to its intervals of 6th and 3rd the interval of a 5th.

It may be produced by taking the first inversion of the chord of the 7th upon the subdominant of some key in the minor mode, and augmenting the subdominant chromatically. Thus, if we take F A C D, which is the first inversion of D F A C, the chord of the 7th upon D the subdominant in A minor, and augment D into D♯, we have F A C D♯, which is a chord of the extreme sharp 6th with a 5th.

The chord so produced is also capable of being resolved upon the dominant of the same key, so as to make a species of imperfect cadence in that key; and the resolution is into a common chord of $\frac{8}{5}$, the bass note and the 3rd and 5th of the 3 bass note each descending a semitone, and the augmented 6th ascending a semitone. But, in order to avoid the consecutive 5ths which such a resolution would involve, composers, in general, think it proper to interpose a chord of $\frac{6}{4}$*, as in the example in the Table, sup. p. 66, No. 37, where the chord F A C D♯ is resolved by the semitonic descent of the bass F into E, and the semitonic ascent of the 6th, D♯, into E, —retaining, in the first instance, the 5th, C, and the 3rd, A, to form a 6th and 4th to the new bass E; after which, the final resolution into the common chord E G♯ B E takes place by the semitonic descent of C and A respectively into B and G♯; and the consecutive 5ths, F C and E B, which would have occurred, if the 6th and 4th had not been interposed, are thus avoided. According to many writers, however†, there is no objection to proceeding at once to the final resolution; thus:

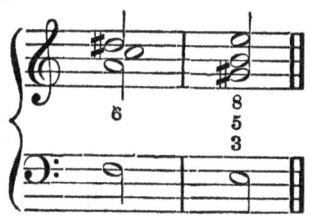

---

* See Callcott's Grammar, p. 241.

† See Albrechtsberger, vol. i, p. 24, note by M. Choron. And see Appendix, Note (AAA).

as they hold the consecutive 5ths, in this case of the resolution of the $\frac{6}{5}$, to be (by exception from the common rule)
3
inoffensive to the ear.

Such are the principal points for remark with regard to
this chord. In other respects, it is subject, in general, to the
same observations as the chord of the extreme sharp 6th,
which it much resembles in point of effect.

The following is an example of this chord :

Ex. 1.   (*12th Mass*) *Mozart.*

We shall give, as another example, a passage in which
this chord is employed with peculiar felicity of effect; and
which, taken altogether, may fairly be considered as one of
the finest conceptions in music.

Ex. 2.   (*Creation*) *Haydn.*

Lord   and King of   Na - ture all.

XXVIII. *The Chord of the Extreme Sharp 6th with a Tritone.*

(V. sup. Table, p. 67, No. 38.)

This chord is of the same strictly extraneous character as the last two; and differs from the chord of the extreme sharp 6th only by adding to its intervals of 6th and 3rd the interval of a tritone.

It may be produced by taking the second inversion of the chord of the 7th upon the 2nd ascending, of some key in the minor mode, and augmenting the subdominant (which is one of the notes of that chord) chromatically. Thus, if we take F A B D, which is the second inversion of B D F A (the chord of the 7th upon B, the second note in A minor), and augment the subdominant D into D♯, we have F A B D♯, which is a chord of the extreme sharp 6th with a tritone. The chord so produced is also capable of being resolved upon the dominant of the same key, so as to make a species of Imperfect Cadence in that key; and the resolution is into a common chord of $\frac{8}{5}$, by the semitonic descent both of the bass note and its 3rd, and the semitonic ascent of the augmented 6th; while the note which formed the tritone to the bass remains. Thus, in the Example in the Table, sup. p. 66, No. 38, the chord

F A B D♯ resolves into E G♯ B E, by F's falling a semitone to E, A's falling a semitone to G♯, and D♯'s rising a semitone to E, while B remains.

This chord is subject, in general, to the same remarks as apply to the last two. It is somewhat harsher, however, than either of these, and is much less frequently used*.

The following specimen of its effect may suffice.

(*Anthem*) *C. P. E. Bach.*

* See Appendix, Note (BBB).

169

XXIX. *The Chord of the Major 7th, Minor 6th, 4th, and 2nd.*
(V. sup. Table, p. 67, No. 39.)

The following is an example of this chord.

Ex. 1.   (*Zauberflöte*) *Mozart.*

XXX. *The Chord of the Diminished 7th, with its Inversions.*
(V. sup. Table, p. 69, Nos. 41, 42, 43.)

This chord always comprises the interval of a diminished minor 7th, and generally also those of a diminished 5th and minor 3rd.

Its bass note may be either the leading note of some key in the minor mode; or the subdominant, chromatically augmented, of some key in the major mode.

In the resolution, its bass note usually ascends a semitone, and the note forming a 7th to the bass usually descends a semitone. Thus, in the Example, sup. p. 116, the chord of the dominant 7th, G♯ F B D, resolves into the common chord minor, A E C, the bass note, G♯, ascending to A, and the 7th, F, descending to E.

This chord has three inversions; the first, figured $\frac{6}{5}$, con-
sists of a major 6th, minor 5th, and a minor 3rd. The second, figured $\frac{6}{4}$, of a major 6th, an augmented 4th, and a minor 3rd.

And the third, figured $\frac{6}{4}$, of a major 6th, tritone, and aug-
2

mented 2nd. The first is produced by taking for the bass note that note which forms the 3rd of the direct chord; the second, by taking for the bass note that note which forms the 5th of the direct chord; the third, by taking for the bass note that note which forms the 7th of the direct chord.

The direct chord and its inversions are all distinguished by the fine and peculiar expression of various kinds which they are capable of imparting to passages in which some particular effect is designed, and also for the services they render in Cadences and Modulations*; and they are in more frequent use than any other of the extraneous discords.

The following are examples of the chord of the diminished 7th.

Ex. 1.

Corfe.

Ex. 2.

(*Adelaide*) *Beethoven.*

---

* Vide sup. pp. 109, 116.

Ex. 3.

(Surprise) Haydn.

Ex. 4.

(*Zauberflöte*) *Mozart.*

Ex. 5.

(*12th Mass*) *Mozart.*

Mi - se - re - re,   mi - se - re - re,

Mi - se - re - re,   mi - se - re - re,

Mi - se - re - re,   mi - - se - re - re,

Mi - se - re - re,   mi - - se - re - re,

The following examples will illustrate the three inversions of the diminished 7th.

Ex. 1.   (*Creation*) *Haydn.*

* First inversion of the chord of the diminished 7th on F♯.

Ex. 2.   *(Elijah) Mendelssohn.*

And they seek my   life,   and they seek my

life,   to   take

* First inversion of the diminished 7th on G♯.
† First inversion of the diminished 7th on A♯.

Ex. 3.

(*Surprise*) *Haydn.*

* First inversion of the diminished 7th on B♯.
† Diminished 7th on B♯.   ‡ Ditto.

Ex. 4.

(Adelaide) Beethoven.

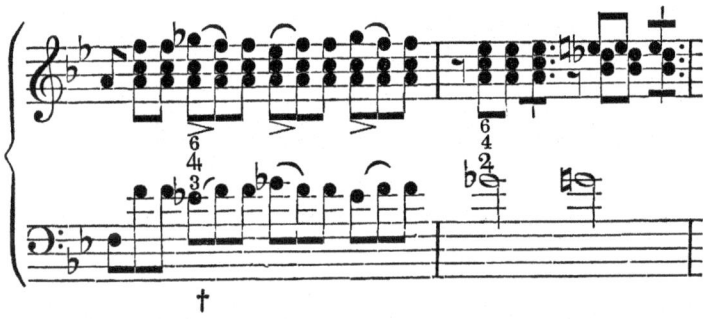

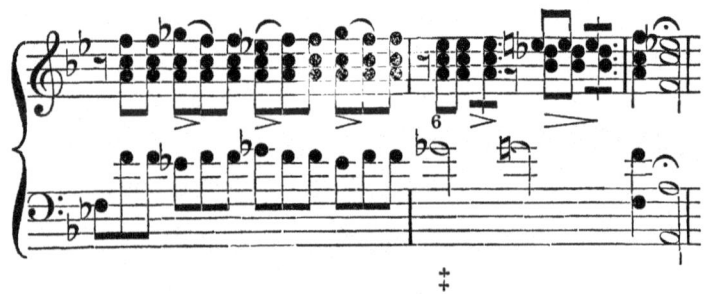

* Second inversion of the chord of the diminished 7th on C♯.
† Second inversion of the diminished 7th on A.
‡ Third inversion of the diminished 7th on A.

Ex. 5.

(Creation) Haydn.

Ex 6.

(Creation) Haydn.

Here   shoots   the heal - ing plant.

---

* Third inversion of the diminished 7th on C♯.

† Third inversion of the diminished 7th on A.

N

CHAPTER V.

THE pitch or elevation of a note depends on the degree o the frequency of the vibrations of the string or other sounding body by which it is produced†.

The frequency of the vibrations in strings increases with their *shortness, lightness,* and *tension.* It is found that a string of one-half the length, or one-fourth the weight, or of four times the tension, of another string, vibrates twice as fast on any of these accounts‡.

According to the musical scale now in use, the numerical expression (fractionally) of the ratios of the vibrations productive of the notes in any key in the major mode, ascending from the fundamental or key-note (suppose C), to the vibrations productive of *that* note, is as follows :

C	D	E	F	G	A	B	C §
1	$\frac{9}{8}$	$\frac{5}{4}$	$\frac{4}{3}$	$\frac{3}{2}$	$\frac{5}{3}$	$\frac{15}{8}$	2

---

* See Appendix, Note (CCC).

† Holder, on Harmony, p. 56 ; Arnott's Physics, vol. i, p. 511.

‡ Arnott's Physics, vol. i, p. 506 ; Whewell's Inductive Sciences, vol. ii, p. 333.

§ See Appendix, Note (DDD).

which makes the ratios of the vibrations as between any two *contiguous* notes as follows :

C to D	D to E	E to F	F to G	G to A	A to B	B to $\bar{C}$*
$\dfrac{9}{8}$	$\dfrac{10}{9}$	$\dfrac{16}{15}$	$\dfrac{9}{8}$	$\dfrac{10}{9}$	$\dfrac{9}{8}$	$\dfrac{16}{15}$

The intervals of $\frac{9}{8}$ are called *major tones;* those of $\frac{10}{9}$ *minor tones;* and those of $\frac{16}{15}$ *semitones;* so that the whole scale in the major mode, which was before stated, generally, and in a way sufficient for practical purposes, to consist of *tone, tone, semitone, tone, tone, tone, semitone*†, consists, if more closely and scientifically considered, of

Major tone, minor tone, semitone, major tone, minor tone, major tone, semitone.

The ratios of the vibrations which produce the notes of the scale in the *minor* mode to the vibrations which produce the key-note, seem to be as follows :

Descending.

$\bar{A}$	G	F	E	D	C	B	A ‡
2	$\dfrac{9}{5}$	$\dfrac{8}{5}$	$\dfrac{3}{2}$	$\dfrac{4}{3}$	$\dfrac{6}{5}$	$\dfrac{9}{8}$	1

Ascending.

A	B	C	D	E	F♯	G♯	A §
1	$\dfrac{9}{8}$	$\dfrac{6}{5}$	$\dfrac{4}{3}$	$\dfrac{3}{2}$	$\dfrac{5}{3}$	$\dfrac{15}{8}$	2

and the ratios as between two *contiguous* notes in the minor mode will be as follows :

Descending.

A to G	G to F	F to E	E to D	D to C	C to B	B to A
$\dfrac{10}{9}$	$\dfrac{9}{8}$	$\dfrac{16}{15}$	$\dfrac{9}{8}$	$\dfrac{10}{9}$	$\dfrac{16}{15}$	$\dfrac{9}{8}$

---

* See Appendix, Note (EEE).   † V. sup. p. 3.
‡ See Appendix, Note (FFF).   § See Appendix, Note (GGG).

Ascending.

A to B	B to C	C to D	D to E	E to F♯	F♯ to G♯	G♯ to A*
$\frac{9}{8}$	$\frac{16}{15}$	$\frac{10}{9}$	$\frac{9}{8}$	$\frac{10}{9}$	$\frac{9}{8}$	$\frac{16}{15}$

that is :

Descending.

Minor tone, major tone, semitone, major tone, minor tone, semitone, major tone.

Ascending.

Major tone, semitone, minor tone, major tone, minor tone, major tone, semitone.

On examination of these ratios, it will be obvious that they are all based on the four lowest prime numbers, 1, 2, 3, 5 ; or, in other words, on the ratios borne to 1, by 2, 3, and 5 ; and that there is not a single instance in which any of the higher prime numbers, such as 7, 11, 13, &c. is concerned.

This is a remarkable fact ; and such as to suggest the conception that the scale might possibly have originated in a course of invention, of which the leading idea was, that of applying the lowest possible prime numbers to the formation of a system of sounds, separated from each other by such distances of pitch as are found suitable to the human voice, and sufficient in number for the purposes of song. It is by no means intended to express an opinion that such was in fact the history of the matter. The evidence, indeed, that exists on the subject tends to the conclusion, that the scale is due to no single idea or single inventor, but was in the nature of a satisfactory and final improvement upon preceding arrangements, by which it had been in part anticipated†. But, however this may be, it will be both useful and curious to endeavour to trace the progress by which the single idea, that has been referred to, *might* in the nature of things have led some particular inventor to this great discovery.

---

* See Note, Appendix (HHH).   † See Note, Appendix (III).

He shall be supposed then to set out by taking a string sounding some particular note, as C. His next step would be to take another string, whose vibrations would be to the vibrations of the first as $2 : 1$; and he would find that this sounded $\overline{C}$, the octave ascending, which his ear would feel as substantially the same sound*, so that the first note would return, as it were, into itself; and though he might go higher or lower by the repeated use of the same numbers, 1 and 2, ascending and descending,—for example, by resorting to strings which should vibrate in proportion to C as $4 : 1$, or as $\frac{1}{2} : 1$, these would still produce only new repetitions or octaves of the first note, yielding substantially the same sound.

To obtain new notes, therefore, he would be obliged to resort to the next lowest prime number, 3, by taking a string which should vibrate in proportion to C as $3 : 1$. This would produce a note in the second octave ascending, viz. the octave commencing with $\overline{C}$; for $\overline{C}$ is to C, as $2 : 1$, and the 8th above it would consequently be as $4 : 1$; and, therefore, $3 : 1$ would fall between these, or, in other words, within the limits of the second octave ascending. But the interval between this note and C, would be too great to suit the voice or ear, so far as ordinary musical purposes are concerned. It would be necessary, therefore, to lower this new note, so as to bring it within the first octave ; and this might be done by the application of number 2, or, in other words, by resorting to a string which should vibrate as $\frac{3}{2} : 1$, which would give him G in the first octave†. If he were now again to employ number 3, by taking another string vibrating three times as often as string 3, viz. as $9 : 1$, it would sound a note in the fourth octave ascending, which, being much too remote from the original note, would require to be lowered; and this might be done by an application of number 2 to the third power, viz. by resorting to a string vibrating $\frac{1}{8}$th as often as string 9 (or as $\frac{9}{8} : 1$); and this would bring the new note

---

* V. sup. p. 2.　　† See the Ratios of the Scale, sup. p. 180.

down within the compass of the first octave, when it would become note D in the first octave.

Deserting now (for the present) number 3, and taking 5 instead*; suppose him to resort to another string vibrating in proportion to C as $5:1$. This string would produce a note in the third octave ascending, which, being lowered (upon the same principle as in the former cases) two octaves by the application of number 2 to the second power, so as to be expressed by $\frac{5}{4}:1$, would become E in the first octave. And if he were to apply to this same string 5, the former number of 3, taking a string which should vibrate in proportion to C as $15:1$, it would produce a note in the fourth octave ascending, which, being lowered by the application of 2 to the third power, so as to be expressed by $\frac{15}{8}:1$, would become B in the first octave.

He would thus have obtained C, D, E, G, B, $\overline{\text{C}}$; and in order to obtain two other notes, such as would divide the large intervals at present left between E and G and between G and B, as well as increase the variety of sounds, he might recur again to the number 3, and take another string, the vibrations of which should be in the *descending* series, viz. as $\frac{1}{3}:1$, so that this string would be exactly as much below string 1, as string 1 was below string 3. This new string would yield a note in the second octave descending, which, being *raised* (upon the same principle that the note was *lowered* in the former cases) two octaves by the application of number 2 to the second power, so as to be expressed by $\frac{4}{3}:1$, would become F in the first octave. And if he were next to apply to this same string of $\frac{1}{3}$ the number 5 (for number 3 would only produce the key-note again, viz. $\frac{3}{3}:1$, and its application would consequently be useless), he would have a note falling within the first octave, and standing in relation to the key-note as $\frac{5}{3}:1$; which note so obtained would be A, and the first octave would consequently be complete.

---

* V. post, p. 187, note *, and the note in the Appendix there referred to.

184

All this process may be summed up in the following Table, which will shew at the same time in what manner the same four prime numbers conduce to the construction of the scale in the *minor* mode.

TABLE,

Shewing how the Ratios of the Simple Intervals formed with the Key-note are founded on the Numbers 1, 2, 3, 5.

Major Mode*.

String vibrating as

1, will produce...... ....Key-note

2............................... ....8th

$\frac{3}{2}$............................. ....5th

$\frac{3\times3}{2\times2\times2}$ $(=\frac{9}{8})$...... ....2nd

$\frac{5}{2\times2}$ $(=\frac{5}{4})$......... ....3rd

$\frac{5\times3}{2\times2\times2}$ $(=\frac{15}{8})$......... ... .7th

$\frac{1}{3}\times2\times2$ $(=\frac{4}{3})$............. . ...4th

$\frac{1}{3}\times5$ $(=\frac{5}{3})$........... ....6th

Minor Mode descending†.

String vibrating as

1, will produce... ....Key-note

2................... . ...8th

$\frac{3}{2}$........... ....5th

$\frac{3\times3}{2\times2\times2}$ $(=\frac{9}{8})$................ . ...2nd

$\frac{1}{3}\times2\times2$ $(=\frac{4}{3})$......... . ...4th

* See the ratios of the scale in the major mode, sup. pp. 179, 180.

† See the ratios of the scale in the minor mode descending and ascending, sup. pp. 180, 181.

$\frac{1}{5} \times 2 \times 2 \times 2 \ (=\frac{8}{5})$...... ....6th

$\frac{1}{5} \times 3 \times 2 \ (=\frac{6}{5})$.............. . ...3rd

$\frac{1}{5} \times 3 \times 3 \ (=\frac{9}{5})$............ . ...7th

Minor Mode ascending.

String vibrating as

    1, will produce. ....Key-note

    2 ....8th

    $\frac{3}{2}$·· 5th

    $\frac{3\times3}{2\times2\times2} \ (=\frac{9}{8})$. 2nd

    $\frac{5\times3}{2\times2\times2} \ (=\frac{15}{8})$..................... ....7th

    $\frac{1}{3} \times 2 \times 2 \ (=\frac{4}{3})$.............. . ...4th

    $\frac{1}{3} \times 5 \ (=\frac{5}{3})$............ ....6th

    $\frac{1}{5} \times 3 \times 2 \ (=\frac{6}{5})$.............. . ...3rd

The consideration of which Table incidentally suggests the following remark, as to the distinction between the major and the minor modes (a remark not contained, it is believed, in any preceding treatise); viz. that it consists in the introduction into the minor mode of the prime number 5, in the *inverted* form of $\frac{1}{5}$, instead of the *direct* form employed in the major*.

Such being the ratios on which the simple intervals of the scale are founded, it remains only to remark that all other musical intervals (whether compound intervals of the scale or extraneous discords) are compounded of these same ratios; and that it is consequently true of these as well as the former,

---

* It will be observed, that 5 in the direct form occurs not at all in the *descending* or *proper* progression of the minor mode, though it occurs both in the direct and inverted form in the *ascending*.

that they are all founded on the same prime numbers, 1, 2, 3,
5. The following Table comprises the chief intervals of all
kinds (including the simple intervals of the scale), and at the
same time exhibits in what manner and in what proportion
the prime numbers in question enter into each interval.

TABLE,

*Shewing how the Ratios of the Musical Intervals are
universally founded on the Numbers 1, 2, 3, 5.*

	Compounded of
A Semitone (or Diatonic Semitone)............... $\frac{16}{15}$	$\frac{2\times2\times2\times2}{5\times3}$
Minor Tone............... $\frac{10}{9}$	$\frac{5\times2}{3\times3}$
Major Tone............... $\frac{9}{8}$	$\frac{3\times3}{2\times2\times2}$
Minor Third . ... $\frac{6}{5}$*	$\frac{3\times2}{5}$
Major Third............... $\frac{5}{4}$	$\frac{5}{2\times2}$
Fourth...................... $\frac{4}{3}$	$\frac{2\times2}{3}$
Tritone............. .... $\frac{45}{32}$	$\frac{3\times3\times5}{2\times2\times2\times2\times2}$

* See Note, Appendix (JJJ).

Minor Fifth............	$\frac{64}{45}$	$\frac{2\times2\times2\times2\times2\times2}{3\times3\times5}$
Fifth.........................	$\frac{3}{2}$	$\frac{3}{2}$
Minor Sixth..............	$\frac{8}{5}$	$\frac{2\times2\times2}{5}$
Major Sixth......	$\frac{5}{3}$*	$\frac{5}{3}$
Minor Seventh............	$\frac{16}{9}$†	$\frac{2\times2\times2\times2}{3\times3\dotsc}$
Major Seventh............	$\frac{15}{8}$	$\frac{3\times5}{2\times2\times2}$
Octave......................	2	2
Minor Ninth	$\frac{32}{45}$	$\frac{2\times2\times2\times2\times2}{3\times3\times5}$
Major Ninth	$\frac{9}{4}$	$\frac{3\times3}{2\times2}$
Minor Tenth	$\frac{12}{5}$	$\frac{2\times2\times3}{5}$
Major Tenth	$\frac{5}{2}$	$\frac{5}{2}$
Eleventh....	$\frac{8}{3}$	$\frac{2\times2\times2}{3}$

* See Note, Appendix (KKK).   † See Note, Appendix (LLL).

Twelfth .... ·	3	3
Minor Thirteenth	$\dfrac{16}{5}$	$\dfrac{2\times2\times2\times2}{5}$
Major Thirteenth.........	$\dfrac{10}{3}$	$\dfrac{5\times2}{3}$
Minor Fourteenth.	$\dfrac{32}{9}$	$\dfrac{2\times2\times2\times2\times2}{3\times3}$
Major Fourteenth.	$\dfrac{15}{4}$	$\dfrac{3\times5}{2\times2}$
Fifteenth...................	4	4
Minor Sixteenth..........	$\dfrac{64}{15}$	$\dfrac{2\times2\times2\times2\times2\times2}{3\times5}$
Major Sixteenth.	$\dfrac{9}{2}$	$\dfrac{3\times3}{2}$
Minor Seventeenth.......	$\dfrac{24}{5}$	$\dfrac{2\times2\times2\times3}{5}$
Major Seventeenth.......	5	5
Enharmonic Interval. ...	$\dfrac{128}{125}$	$\dfrac{2\times2\times2\times2\times2\times2\times2}{5\times5\times5}$
Chromatic Semitone	$\dfrac{25}{24}$*	$\dfrac{5\times5}{2\times2\times2\times3}$

* See Note, Appendix (MMM).

Augmented Second...... $\frac{75}{64}$	$\dfrac{3\times5\times5}{2\times2\times2\times2\times2\times2}$
Diminished Minor Third $\frac{144}{125}$	$\dfrac{3\times3\times2\times2\times2\times2}{5\times5\times5}$
Augmented Major Third $\frac{125}{96}$	$\dfrac{5\times5\times5}{3\times2\times2\times2\times2\times2}$
Diminished Fourth ...... $\frac{32}{25}$	$\dfrac{2\times2\times2\times2\times2}{5\times5}$
Augmented Fourth...... $\frac{25}{18}$	$\dfrac{5\times5}{3\times3\times2}$
Diminished Fifth......... $\frac{36}{25}$	$\dfrac{3\times3\times2\times2}{5\times5}$
Augmented Fifth......... $\frac{25}{16}$	$\dfrac{5\times5}{2\times2\times2\times2}$
Diminished Minor Sixth $\frac{192}{125}$	$\dfrac{3\times2\times2\times2\times2\times2\times2}{5\times5\times5}$
Augmented Major Sixth (or Extreme Sharp Sixth). .... $\frac{125}{72}$	$\dfrac{5\times5\times5}{2\times2\times2\times3\times3}$
Diminished Minor Seventh . ...$\frac{128}{75}$*	$\dfrac{2\times2\times2\times2\times2\times2\times2}{5\times5\times3}$

* See Note, Appendix (NNN).

TEMPERAMENT.

It appears, from what has been already stated*, that the *tones* of the scale are unequal, some being major tones, of the ratio of 9 : 8, and some minor tones, of the ratio of 10 : 9. The difference between these is 81 : 80, called, in scientific disquisitions, a *comma*.

This difference between the major and minor tones would make it necessary, in the series of keys, supposing the scale to be exactly adhered to, to introduce another new note in addition to the new note (sharp or flat) which, as formerly explained†, is now in fact always introduced at each successive key. Thus, as, in proceeding from the key of C major to that of G major, we now always introduce the new note F♯, in order to preserve in the new key the right relative arrangement of tones and semitones, so it would be requisite to introduce another new note, if we wished to preserve the right relative arrangement of major tones and minor tones. For, first, the interval from G to A, in the new key of G, being the first interval, ought to be a major tone, answering to the first interval C D in the original key‡; whereas, the interval from G to A in the original key, being the fifth interval, is a *minor* tone. And again, the interval from A to B in the new key, being the second interval, ought to be a *minor* tone, answering to the second interval D E in the original key§; whereas, the interval from A to B in the original key, being the sixth interval, is a *major* tone. In order, therefore, to obviate the errors that would thus be introduced into the key of G, we should be obliged to employ for that key a new note in lieu of A; and if a note were taken for the purpose, sharper than A by exactly a *comma*, it would obviate both errors at once, and make the tune of the new key perfect; for the interval from G to A

---

* V. sup. pp. 180, 181.   † V. sup. p. 7.   ‡ V. sup. p. 180.   § V. ibid.

would thus be enlarged into a major tone, and the interval from A to B of course reduced to the same extent; that is, reduced into a minor tone. So, in every succeeding key with sharps, the second note would consist of the sixth of the preceding key sharpened by a *comma*; and, upon the same principle, it will be found, conversely, that in every key with flats the sixth note would consist of the second of the preceding key flattened by a *comma*. The result, therefore, would be the introduction into the whole system of keys (taken as fifteen in each mode), of no less than twenty-eight new notes, in addition to those which they now comprise ; that is, one for every key into which a sharp or flat enters.

Such new notes are capable, on some instruments (such as the violin), of being made by a slight change in the place of the finger*; but on instruments with fixed keys, as they are called, that is, instruments with finger notes, like the organ or pianoforte, where the sound of every note is fixed by their very construction, so as to be incapable of any (even the least) modification of pitch, in the act of playing, these new notes could only be produced by the inconvenient addition of as many finger notes, or by some other change of construction equally objectionable. In all such instruments, therefore, the contrivance is adopted known in the treatises by the name of *Temperament;* which consists in so tuning the instrument as to abate or *temper* the difference between the major and minor tones throughout the scale and throughout the different keys; or, in other words, to bring all the tones nearly to equality, by reducing in some degree the major tone, and enlarging in the same degree the minor. The inequality that remains being too small to be appreciated by the ear, the tune throughout is thus rendered sufficiently exact. Thus—to recur to our example—C D, and D E, G A, and A B, being all rendered so nearly equal that the ear accepts them as such, C D and D E may now well stand as the first and second intervals in the key of C ; and G A and

---

* See Note, Appendix (OOO).

A B as the first and second in the key of G ; and in proceeding from the former key to the latter, no alteration of the A of the former is required.

This abatement of the inequality between the major and minor tones is (according to the usual plan of temperament) as follows. The instrument is tuned by taking successive *fifths* from a note selected as the pitch note; for this is the method found to be attended with the least difficulty to the ear. But these fifths cannot all be taken exactly true, even for the purpose of tuning in a single key ; for it results, from the nature of the musical ratios, that a continued series of perfect fifths produces major thirds intolerably sharp*. It becomes necessary, therefore, to flatten a little some of these fifths ; and by the expedient of flattening them all to a proper extent, throughout the series, the effect is produced of not only preventing this harshness of the major thirds, but also of bringing the major and minor tones into an approximate equality. According to this plan of temperament, the octave (being the interval in which the ear least tolerates imperfections) is tuned exactly true ; but the other intervals are all to a certain extent incorrect; for even the major thirds, though so far reduced as to prevent harshness, are left rather sharp. The incorrectness, however, is in no instance of such an amount as to occasion dissatisfaction to the ear†.

---

* See Note, Appendix (PPP).
† See Note, Appendix (QQQ).

CHAPTER VI.

WE will proceed now to make some attempt towards the solution of the curious and interesting question, what it is that forms the source of the pleasure which the ear derives from music in general. We say music *in general*, because the proposed investigation will have no reference to the effects which belong to *particular pieces or passages* of music ; though, as to these, we may take the opportunity of remarking summarily, that the pleasure they afford seems to arise in great part from the constant reference to some given key, and the skilful departure from and return to it (suggesting probably to the mind the ideas of arrangement and contrivance) ; and from *imitative* artifices, exciting ideas of a pleasing or impressive character, by some sort of resemblance which the sounds bear to particular scenes, actions, or objects of that character,—and by the ingenuity which the imitation displays. But our present inquiry is of a more abstract kind, and relates to the pleasure derived from musical sounds in general, as opposed to sounds or connexions of sound that are unmusical, or intrinsically disagreeable to the ear. And, in the course of our remarks, we shall advert, first, to the case of a musical sound individually considered ; and then to the case of several

O

musical sounds heard in connexion with each other, whether connected in the way of succession or of contemporaneous combination.

1. With respect to a musical sound, individually considered, it is clearly established as one of its necessary conditions that the vibrations of the string, or other sonorous body producing it, should in point of frequency range within certain limits ; the sound becoming unmusical when the vibrations, as to their number in a given time, exceed one of these limits, or fall below the other. What these limits are, may perhaps be not fully settled*; and it is possible that they may vary in some degree with different ears.

As to this condition of a musical sound (the degree of frequency in the vibrations), the reason on which it rests is somewhat obscure ; and the speculation upon it seems to involve the doctrine of vibration in general, and to raise the question why and in what manner it is a necessary agent (as we know it in fact to be) in producing the perception both of sound and light. We may conjecture, however, that when the vibrations reach a certain degree of frequency, their rapidity confounds the ear, and prevents their separate character from being perceived; and that so, on the other hand, when they fall below a certain degree of frequency, the slowness of their succession prevents the ear from connecting them together. But the condition to which we have referred is not the only one that belongs to a musical sound. It is also essential to it, that the vibrations of the body producing it should be uniform, that is, that the number should be equal in equal times; as to which condition, we may add this remark, that, in order to secure its observance, the string or other sounding body should itself be of regular or symmetrical shape.

2. In proceeding to inquire into the second point, viz. the source of the pleasure derived from several musical sounds heard in connexion with each other, our attention will, at the outset, be due to the following facts.

---

* Vide sup. p. 3, Note †.

1st. *The ratios of the musical intervals are based exclusively on the four lowest prime numbers, 1, 2, 3, 5.*

This has been abundantly established and illustrated in the course of the last chapter.

2nd. *The ratios of the concordant intervals are composed of the numbers 1, 2, 3, 5, in their primary form, or multiplied only by 2, or by some power of 2 ; but the ratios of the discordant intervals always comprise the product of the higher numbers 3 or 5, multiplied into themselves respectively, or multiplied into each other.*

This remarkable fact, which the author does not recollect to have seen pointed out in any other work, will be manifest in the face of the last Table. Thus the ratio of the concordant interval of a minor 3rd (viz. $\frac{6}{5}$) is composed of 3 multiplied by 2 and divided by 5 ; and that of the concordant interval of the major 3rd (viz. $\frac{5}{4}$), of 5 divided by 2×2 ; but the ratio of the discordant interval of a semitone (viz. $\frac{16}{15}$) is composed of 2 multiplied into itself four times, and divided by 5 multiplied into 3 ; and the ratio of the discordant interval of a minor tone (viz. $\frac{10}{9}$) is composed of 5 multiplied by 2 and divided by 3×3. And so of all the other concordant and discordant intervals.

3rdly. *Of the discordant intervals it may be said, in general, that those are harsher whose ratios comprise a multiplication by a higher number, or by a higher power of the same number ; for example, a multiplication by 5, as compared with a multiplication by no higher number than 2 or 3 ; or a multiplication by 5 × 5, as compared with a multiplication by 5, or by 5 × 3.*

Thus the *extraneous* intervals, which (as appears by the last Table) universally involve in their ratios the multiplication of 5 × 5 (extending sometimes to 5 × 5 × 5), are also generally harsher than a major tone, or a minor 7th, or a

major 9th, which involve in their ratios no higher multiplication than 3×3; or even than a major 7th, whose ratio
comprises none higher than 5×3. And thus, too, the major
7th is, in its turn, harsher than the minor 7th, whose ratio
comprises no higher multiplication than 3×3.

4thly. *Of the discordant intervals it may also be said, in
general, that those are harsher whose ratios are of less
magnitude.*

Thus the semitone, having a ratio of $16 : 15$, which is less
than $45 : 32$ or $64 : 45$, the ratios of the tritone and minor 5th
respectively, is also harsher than the intervals last mentioned;
and even the major tone, the ratio of which is $9 : 8$, is, for
the same reason, harsher than these, though its ratio comprises the lower multiplication only of 3×3, while that of the
tritone and minor 5th comprises $3 \times 3 \times 5$.

When these facts are carefully considered, they seem to
warrant the conclusion, that that constitution of musical sounds
by which all their ratios are in fact based upon the elementary numbers of 1, 2, 3, 5, is so far at least perceived by the
ear, in listening to the sounds, as to affect it with the sense
of a certain order and proportion, mingled with a sufficient
variety; which sense is the source of the pleasure derived
from musical sounds heard in connection with each other[*];
that this sense is less distinct when the ratios comprise a
multiplication by a higher number; or a more frequent
multiplication by the same number; or are of less magnitude; because in any of these cases they are the less
clearly apprehended; and that this is the reason why discords are less pleasing than concords, and some discords
less pleasing than others, or even absolutely displeasing; it
having been shewn, as a matter of fact, that the whole class
of discordant *intervals* are distinguished from such as are
concordant, by comprising in their ratios a multiplication of

[*] See Note, Appendix (RRR).

the two highest numbers into themselves, or into each other,—
which in the concordant intervals never occurs; and that
among the discordant intervals those are the harsher whose
ratios comprise a multiplication by a higher number; or a
more frequent multiplication by the same number; or are of
less magnitude. It is true, that the case of *chords comprising
several intervals* is more complex, and that their character is
affected by a greater variety of circumstances; but it will be
found true as to these also, that the ratios of all such as are
concords are exclusively composed of the numbers 1, 2, 3, 5,
either in their primary form, or multiplied only by 2, or some
power of 2; and that among those which are discords, the
comparative harshness mainly depends upon that of the intervals which they comprise, and particularly of those which are
formed between their *extreme* notes. Thus the common
chord, C E G $\overline{\text{C}}$ (which is a concord), has the following in-
tervals with the following ratios: $\text{C E} = \frac{5}{4} = \frac{5}{2 \times 2}$, $\text{C G} = \frac{3}{2}$

$\text{C }\overline{\text{C}} = 2$, $\text{E G} = \frac{6}{5} = \frac{2 \times 3}{5}$, $\text{E }\overline{\text{C}} = \frac{8}{5} = \frac{2 \times 2 \times 2}{5}$, $\text{G }\overline{\text{C}} = \frac{4}{3} = \frac{2 \times 2}{3}$
and in all these no multiplication of 3×3, or 3×5, or 5×5 oc-
curs. On the other hand, the chord of the 9th, C E G $\overline{\text{D}}$ (which
is a discord), has the following intervals with the following ra-
tios: $\text{C E} = \frac{5}{4} = \frac{5}{2 \times 2}$, $\text{C G} = \frac{3}{2}$, $\text{C }\overline{\text{D}} = \frac{9}{4} = \frac{3 \times 3}{2 \times 2}$, $\text{E G} = \frac{6}{5} = \frac{3 \times 2}{5}$

$\text{E }\overline{\text{D}} = \frac{9}{5} = \frac{3 \times 3}{5}$, $\text{G }\overline{\text{D}} = \frac{3}{2}$; and in this the multiplication by
3×3 occurs. And so the chord of the 7th, C E G B (which
is a discord), has the following intervals with the follow-
ing ratios: $\text{C E} = \frac{5}{4} = \frac{5}{2 \times 2}$, $\text{C G} = \frac{3}{2}$, $\text{C B} = \frac{15}{8} = \frac{3 \times 5}{2 \times 2 \times 2}$

$\text{E G} = \frac{6}{5} = \frac{3 \times 2}{5}$, $\text{E B} = \frac{3}{2}$, $\text{G B} = \frac{5}{4}$; and in this the multiplication by 3×5 occurs. And of these discords the latter
is the harsher; the reason of which is, that the interval C B,
formed between its extreme notes, is, according to the rules
above laid down, harsher than the interval C $\overline{\text{D}}$, formed
between the extreme notes of the former discord.

That the pleasure in question has here been referred to
its true source, will appear the more probable, if we consider

the analogous effect on the eye, when it is affected with the sense of order and proportion, mingled with due variety, in architecture, or other productions of art. To inquire further, and to attempt to investigate the principle on which this sense should in either case be productive of pleasure, would be to enter upon speculations beyond the design of this work. But we will hazard the suggestion, that, so far as order and proportion at least are concerned, it may be referred, with some probability, to the idea, that they naturally convey, of *design*, and may thus be traced to a moral and intellectual source.

The views, however, that have been entertained by theorists on the subject before us, particularly as regards the distinction between concords and discords, differ considerably from those above suggested.

The reason most usually assigned for the difference of effect that belongs to concords and discords, is, that, the intervals of which the former consist involving simpler ratios of vibrations, there will consequently be a more frequent *coincidence* of vibrations ; or, in other words, the sounding bodies will set out on their courses together more frequently : the effect of which will be to occasion less jarring and collision, and to produce more unity, fulness, and smoothness of sound*. But this theory has been powerfully assailed by a celebrated writer†, who objects, in the first place, that there will be *no* coincidence of vibrations (however simple their ratios), unless we suppose that the several sounding bodies begin to vibrate precisely at the same moment; and further, that the ratios themselves of the concords are almost all of them *altered*, on instruments with fixed keys, by the effect of *temperament*; and yet that the character of the concords is not thereby sensibly impaired. He also argues, that if bodies, vibrating in a certain ratio to each other, yield a concord on account of the coincidence of their vibrations, we should naturally expect that bodies vibrating

---

* See Note, Appendix (SSS.) † See Note, Appendix (TTT).

in a proportion not very different, would also yield a concord, though one somewhat less harmonious than the other; but that the fact is not so—the ratio of 6 : 5, by example, which is that of the minor third, producing a very agreeable concord, while that of 7 : 6 is wholly inadmissible, and shocking to the ear.

The objector, however, while he combats with so much force the theory in question, is unable to offer a better one—the view that he suggests on the subject to which it relates, amounting to little more than a confession of ignorance. The concords (as he truly remarks) are all to be found in the common chord, or its inversions; which contains, on the other hand, no discord. It is natural therefore (he argues) that the concords, and these alone, should partake of the harmonious quality which belongs to the common chord, their parent; and if the question should still be proposed, why the common chord itself should give so much pleasure, while other sounds offend the ear, what answer (he inquires) can be given, except by asking, in return, why green is more agreeable to the eye than grey ?

We recur then to the doctrine above advanced; viz. that which accounts for concords by the greater distinctness with which their ratios (owing to their greater simplicity) are apprehended, and the clearer sense of order and proportion which is consequently produced—a doctrine which seems liable to none of the objections advanced against the other.

At first sight, indeed, it may appear that the argument derived from *temperament* applies equally to both; but if it be applied to that founded on the more distinct perception of the ratios, it seems a sufficient answer, that a very small alteration of the ratios does not suffice to prevent us from receiving the general impression of them; that our sense of them is not so nice, but that the ear is content to take the *tempered* interval as the true one; and that, though it tolerates no such imperfection in the case of the *octave* (which *temperament* is consequently obliged to leave

untouched)*, that is because the ratio of 2 : 1 is so particularly simple as to cause it to be discerned with more precision ;
and therefore to render the least departure from it painfully
observable.

It may be thought, too, that the objection founded on the
consideration that an interval involving a given ratio shall be
concordant, when that involving a ratio not very different
shall be not merely less concordant, but harsh, or even inadmissible, applies to both doctrines alike. But, as regards that
founded on the greater simplicity and more distinct perception of the ratios, a sufficient answer seems to be afforded by
the fact (not noticed, perhaps, in any other work), that the
concords are all founded on the simple proportions formed by
the primary numbers 1, 2, 3, and 5, multiplied by 2 only, or
some power of 2 ; while the discords, on the other hand, are
founded on proportions formed by 3 and 5, multiplied into
themselves, or each other ; so that there is not only a difference of *degree*, but a sort of *generic* difference, between concords and discords in respect to the relative simplicity of their
ratios ; which may serve sufficiently to account for the marked
line of distinction which the ear draws between concords and
discords. At the same time, it is not true (as the objection
seems to suppose) that such of the concords and discords as
stand near to each other in point of simplicity of ratio, are
nevertheless as widely sundered in character as if their ratios
were more unlike. An instance to the contrary may be
noticed in the case of the minor 7th, the ratio of which (in
some cases being 16 : 9—in others, 9 : 5) is not much less
simple than that of 8 : 5, the ratio of the minor 6th; and
which, though generally ranked as a discord (as it ought,
according to our theory, to be), because founded on a ratio
in which 3×3 enters (viz. in the one case, $2 \times 2 \times 2 \times 2 : 3 \times 3$,
and in the other, $3 \times 3 : 5$), is yet a discord of so mild a cha-
racter that it has been sometimes considered as in the nature

* Vide sup. p. 196.

of a concord. And this is what the doctrine that makes concords depend on the sense of the simplicity of their ratios
would lead us to expect; for it would naturally follow from
this, that, even as between the two classes of concords and
discords, those near the boundary line would have some ap-
proximation to each other in point of effect.

Another difficulty, however, presents itself; viz. that of
conceiving how the ear should be capable of forming any
appreciation of the ratio that exists between the vibrations
of bodies emitting different musical sounds, more particularly when the great rapidity of such vibrations is taken
into account. But it is clear that our senses have a certain intuitive power to perceive order and proportion; as
in the example of architecture, where the eye receives at
first sight immediate impression of the symmetry of an edi-
fice, though there has been no opportunity of measuring,
or maturely considering the relative dimensions of the different parts. And as regards the ear in particular, and the
impression it receives of order and proportion in musical
sound, there is a fact closely illustrative of the intuitive faculty
to which we refer; viz. that, in the performance of a piece of
music, every musician has a clear and accurate sense of the
time to which it is composed, and of the several changes
which the time may undergo, and the degree of regularity with
which it is observed; even though he make no resort to beating or counting, to ascertain its rate or correctness. It is
remarkable, too, that the same elementary numbers are concerned in the case of time, as in that of harmonic intervals;
viz. 2, 3, and 5. For the ear will tolerate no time in music
but that of 2, 3 (or, at most, 5) crotchets, quavers, &c. (or
their multiples) in a bar. The limitation is, indeed, somewhat stricter than in the harmonic intervals; for the time of
5 (though practicable) is of such unusual occurrence, that
the three former numbers alone may be said universally to
prevail*.

---

* See Note, Appendix (YYY).

Upon the whole, then, no sufficient reason occurs for doubting the correctness of the theory above maintained—that the charm of music, considered in reference to the relations between the sounds, fundamentally depends on the pleasure which the ear derives from the sense of order and proportion (mingled with sufficient variety), which ratios uniformly founded on the four lowest prime numbers have a tendency to produce. The doctrine may not be without its difficulties; but there is, at least, no other which seems more entitled to reception; and it is believed that increasing observation and reflection will gradually build up the entire demonstration of its truth.

APPENDIX.

(A), p. 3. Any particular sound required may be fixed by a metallic instrument called a tuning fork, the vibrations of which, on being struck, will produce that sound. Thus, a tuning fork of the proper size and construction will produce that sound which musicians agree to call note A or note C; and a tuning fork of that size and construction consequently becomes the universal standard of that note.

(B), p. 3. " The lowest note, which is perceptible to the human ear, has about 30 beats in a second; and the highest, about 30,000; and there is included between these two, a range of nearly ten octaves. To certain ears, the extremes of this range are totally inaudible, as if their power did not reach so far." (Arnott's Elements of Physics, vol. i, p. 512, 5th edition.) On the range upward, however, there exists great variety of estimate. In Euler's Letters, vol. i, p. 14, of the English translation, it is said—" It would appear that we are incapable of determining either the sound of a string which makes less than 30 vibrations in a second, because it is too low, or else of a string which makes more than 7,552 in a second, because it is too high;" and in Schoedler's Book of Nature, p. 51 of the English translation—"that there exist high notes, the vibrations of which number 48,000 in a second." All these statements, besides, seem to relate to the limits between which a sound is *distinctly audible,* rather than to those between which they are *musical;* which would of course be a narrower range.

(C), p. 6. The Treble clef places the note G on the 2nd line of the staff, and makes every note one 13th higher than if it stood on the same line or space in the Bass clef.

The Alto (or Counter Tenor) clef places the note C on the 3rd line of the staff, and makes every note one 7th higher than if it stood on the same line or space in the Bass clef.

The Tenor clef places the same note C on the 4th line of the staff, and makes every note one 5th higher than if it stood on the same line or space in the Bass clef. The following is an example of music set in these four clefs together.

There is also (among others used in the old music) the Soprano clef,

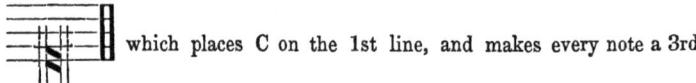 which places C on the 1st line, and makes every note a 3rd

lower than if it stood on the same line or space in the Treble clef.

(D), p. 6. The usual compass or range of notes belonging to Bass, Tenor, Alto, and Treble voices, may be exhibited as follows:

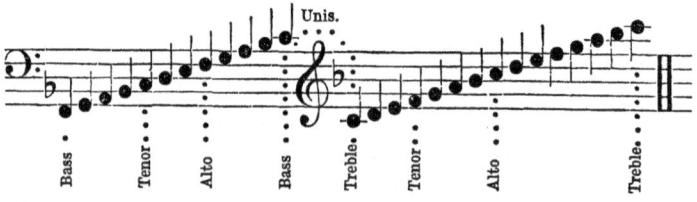

But, in *instrumental* music, the range of the Bass downwards, and of the Treble upwards, is of course considerably greater. The Bass and Tenor voices are those of men; the Treble, of women; and the Alto voices may be found in either sex. There is also a compass called *Mezzo Soprano*, not reaching quite to the farthest extent upwards of the Treble; and a compass called *Contralto*, reaching somewhat lower than the farthest extent downwards of the Alto.

(E), p. 12. Considered in one point of view, the number of keys is 21 in each mode (42 in all) ; for within the octave there are 7 notes ; each of which may be taken in three different ways—viz. as natural, as sharp, and as flat ; and, taken in each way, may be made a key-note. Considered in another point of view, the number of keys is 15 in each mode (30 in all) ; for within the octave in each mode there is one key with the natural signature, 7 with sharp signatures, and 7 with flat ; and this last arrangement is sufficiently extensive ; for the keys which it omits, viz. D♯♯, E♯♯, G♯♯, A♯♯, B♯♯, F♭ major, and B♯♯, E♯♯, C♭, D♭, F♭, G♭ minor, cannot be expressed without double sharps or double flats, which are inconvenient ; a remark, indeed, that applies even to some of the keys that this arrangement includes, viz. the minor keys of G♯, D♯, and A♯, into all of which double sharps enter. Considered, lastly, in another point of view, the number of keys is 12 in each mode (24 in all) ; for not only all the 15 keys in each mode, but all the 21, are capable of being substantially expressed by 12. This will be evident from the consideration that two sounds separated only by an enharmonic interval are substantially identical ; that C♯, for example, is in effect the same with D♭. This being the case, one key may stand for two. Thus the key of D♭ may stand also for that of C♯ ; and, upon the same principle, it will be found that 8 other pairs of keys may be represented by 8 single keys ; which reduces the 21 keys to 12. And it is upon this last arrangement that the construction of instruments with fixed finger-notes or frets, like the organ, pianoforte, or guitar, is founded ; for these finger-notes or frets never exceed the number of 12 within the limits of the octave.

(F), p. 13. This change in the 6th and 7th notes is expressed, in some instances, by a ♯ ; in others, by a ♮. (See, as to these marks, p. 7.) Thus, in the key of G minor, the series is G, A, B♭, C, D, E♮, F, G ; or (which is the same thing) A, B♭, C, D, E, F, G ; for the E in this key would, according to the proper or descending progression of the scale, be *flat*, and therefore the addition of the semitone makes it *natural*.

(G), p. 13. What is here said as to the 6th note, applies more particularly to the scale taken in its entire series. For in *passages*, whether ascending or descending, the 6th note is often taken without the augmentation. (Albrechtsberger, vol. i, p. 93.) The reason for the augmentation in the 7th note seems to be, that without this expedient the key remains undecided. Thus, in a melody comprising only the notes A, B, C, D, E, F, G, A̅, the ear is left in uncertainty whether the key be A minor or C major ; for all the notes equally belong to both keys. But by introducing G♯, the key of A is suggested ; because G♯, on the one hand, is foreign to the key of C, and, on the other, is the *leading* note in the key of A—that is, the note which immediately precedes and therefore naturally leads the ear to A̅, the octave of the tonic ; and the ear, perceiving, at the same time, that the note C is natural, and not sharp,—or, in other words, that the 2nd interval ascend-

ing from the tonic A is B C, a semitone, and not B C♯, a tone (a circumstance characteristic of the *minor* mode—v. sup. p. 2),—collects also that the
key is A minor, and not A major.

(H), p. 21.   This, it will be observed, is *common* time, according to the
definition above given (p. 18); for the number of notes in the bar (6 or 12) is
divisible by 2 ; and again it is *compound*, because the bar in $\frac{6}{4}$ time may be
considered as made up of two bars of $\frac{3}{4}$ time ; in $\frac{6}{8}$ time, as made up of two
bars of $\frac{3}{8}$ time ; and in $\frac{12}{8}$ time, as made up of 4 bars of $\frac{3}{8}$ time.

(I), p. 21.   This is *triple* time, according to the definition above given
(p. 18) ; for the number of notes in a bar—viz. 9—is divisible by 3, and not
by 2.   Again it is *compound*, because the bar in $\frac{9}{4}$ time may be considered
as made up of a bar and a half of $\frac{6}{4}$ time ; in $\frac{9}{8}$ time, as made up of a bar
and a half of $\frac{6}{8}$ time ; and in $\frac{9}{16}$ time, as made up of a bar and a half of $\frac{3}{8}$
time, or of $\frac{6}{16}$ time, which is euqivalent.

(K), p. 23.   The terms *tonic*, *dominant*, and *subdominant*, are of
French origin.   The French harmonists called the first or key note *tonic*,
because it is the foundation of the key *(ton)* ; the 5th ascending, the *domi-*
nant, because, in the regular form of a full cadence (v. sup. p. 197), its
harmony (that is, either its common chord, or chord of the 7th) immediately
precedes, and may therefore be said to *lead* or *govern* the common chord of
the tonic ; and the 4th ascending, the *subdominant*, because, being equivalent to its lower octave, the 5th descending, it is a 5th below its tonic, as
the dominant is a 5th above it.   As to the term *leading note*, applied to the
7th ascending, it has been adopted because, in the melody of the key, the
7th ascending is naturally followed by, or *leads* the octave of the tonic, and
therefore, when sounded, suggests the key itself.   It is called, by the French,
la note sensible, in reference to the same property.   Harmonic denominations of the same kind are also frequently applied to the other notes of the
scale.   Thus the 2nd is called, in many treatises, the *supertonic*, the 3rd
the *mediant*, the 6th the *submediant*.   But these are in less general use,
and of questionable utility.

(L), p. 24.   A minor 3rd is otherwise called a *flat* 3rd ; a tritone, a
sharp 4th ; a minor 5th, a *flat* or *false* 5th ; a minor 7th, a *flat* 7th ; and a
major 7th, a *sharp* 7th.   But these terms of *flat*, *sharp*, and *false*, have a
tendency to convey erroneous ideas as to the nature of the interval described.
The nomenclature in the text is obviously far preferable, and is that, in fact,
now generally adopted by the best writers.   Some of these, indeed, give the
4th the appellation of a *true* 4th, and the 5th the appellation of a *true* 5th,
to distinguish them from a tritone and minor 5th respectively.   But it is
better to speak of 4ths and 5ths generally, and to suppose them the true
intervals intended, unless the contrary be expressed.

(M), p. 28.   The *diatonic semitone* is so called because the scale itself
to which it belongs, and which consists of tones and of semitones of this

kind, is sometimes called the *diatonic scale*, to distinguish it from other scales. For scales of a nature different from this, appear to have been in use in ancient Greece, and to be now in use in some eastern countries. No other scale than this, however, exists in modern Europe; for though passages occur in which intervals of different kind from tones or semitones, viz. chromatic semitones and enharmonic intervals, are introduced, for the sake of a particular effect, no entire piece is ever composed consisting wholly of any intervals except tones and diatonic semitones.

(N), p. 28. These are commonly described, in the treatises, not as " extraneous intervals," but as " altered intervals," *intervalles altérés* (see Traité de l'Harmonie, par M. Fétis), or as " augmented or diminished intervals." Rousseau, in his Dict. de Musique, published in 1768, speaks of them (Art. Relations) as designated by the term of "*fausses relations*," and proceeds to say, that, with the exception of the extreme sharp 6th, they were not admitted into harmony. " On peut pourtant," he adds " les y faire entendre pourvu qu'un des deux sons qui forment la fausse relation, ne sont admis que comme *note de gout*, et non comme partie constitutive de l'accord." At the present day, the greater part of them are, by all composers, freely employed without any limitation of this kind; and some of them at least, viz. the diminished 7th and extreme sharp 6th, were so employed in the music of Italy and Germany long before the period of Rousseau's work. Their nature, however, is very scantily elucidated in the treatises. That they are discords is certain ; and that, intrinsically considered, they are even discords of a harsh character ; and yet it is equally certain that, on instruments with fixed keys, and consequently tuned by *temperament*, an extraneous interval is, in some instances, represented by the same sounds as a concordant one ; and, in every instance, by the same sounds as some interval of the scale, either concordant or discordant. Thus the augmented 2nd, C D♯, has, on the instruments in question, the identical sounds of the minor 3rd, C E♭, which is a concordant interval ; and the extreme sharp 6th, F D♯, the identical sounds of the minor 7th, F E♭, which is a discord of the scale. However, C D♯ is not a concord, and really differs from C E♭ by an enharmonic interval ; and F D♯ is not a discord of the scale, and really differs from F E♭ by the like interval. This difference between any extraneous interval and that interval of the scale which, on an instrument with fixed keys, is expressed by the same sounds — for example, between C D♯ and C E♭, or between F D♯ and F E♭—is accordingly perceived by the ear, if the two intervals be sounded in succession upon an instrument tuned absolutely true ; and (what is very curious) it is perceptible even upon an instrument with fixed keys (notwithstanding the actual identity, in that case, of the sounds), if the course of the harmony be such as to indicate that an extraneous interval, and not an interval of the scale—for example, C D♯, and not C E♭, or F D♯, and not F E♭,—is intended.

Some remarks on the nature of extraneous intervals (intervalles altérés) are to be found in the learned and acute works of M. Fétis, above cited; in the course of which he notices their identity of sounds with particular intervals of the scale. But he speaks of these intervals of the scale as concordant only, without noticing the fact that they are sometimes discords; and he also speaks of this identity of sound as if it were actual, instead of arising, as it really does, from the imperfection of instruments with fixed keys. He attempts, too, to account for the distinction which the ear perceives between an extraneous interval and the concordant one expressed by the same sounds, in a manner not calculated, in the judgment of the present author, to throw

any true light on the subject. " La seconde augmentée

sonnerait à l'oreille comme la tierce mineure 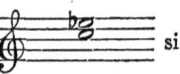 si la succes-

sion des harmonies ne faisait pressentir divers modes de *résolution*, qui jettent l'esprit dans l'incertitude jusqu'à ce que l'une des résolutions soit opérée."

(O), p. 28. Among these, the diminished minor 3rd, the augmented major 3rd, and the diminished minor 6th, are of very rare occurrence. In the table given by M. Fétis, of the intervalles altérés, he introduces also the *diminished octave* and the *augmented octave;* but it is questionable whether these can properly be ranked among musical intervals. On the other hand, he omits the *augmented* 4th and the *diminished* 5th, considering them, perhaps, as identical with the *tritone* and *minor* 5th. But there is, in fact, no such identity, except upon instruments with fixed keys; the ratios of all these intervals being different, as shown in the Table of ratios contained in the present work, sup. p. 186.

(P), p. 30. The case of the chord of the diminished 7th (as to which, v. sup. Table, p. 69, No. 40) involves a sort of exception to this statement. For there, though the bass note be changed, yet, by the aid of an enharmonic transition (as to which, v. sup. p. 118), the intervals and denomination may remain the same. Thus, in the chord of the diminished 7th, C♯, E, G B♭, if E instead of C♯ be made the bass note, and the C♯ enharmonically changed to D♭, the chord of E, G, B♭, D♭, will be produced, which is still a chord of the diminished 7th. But when no enharmonic transition takes place, this chord is subject, like all other chords, to the remark, that a change of the bass note alters the intervals and denomination of the chord, as shown by the Table above referred to, Nos. 41, 42, 43.

(Q), p. 31. This chord is called, by German and latterly by many English writers, the *triad*. But the correctness of the appellation may reasonably be questioned; for it is not (properly speaking) a chord of three notes, but of four; viz. 1st, 3rd, 5th, and 8th. The appellation of triad throws out of the account the 8th, and does so upon the ground, that, being a mere replicate of the bass note, it is substantially the same sound. Yet, in fact, it forms an important element in the chord, to the more full and complete effect of which the ear feels it to be essential, and any appellation which tends to exclude it is consequently defective. The French term of *accord parfait* is not open to this objection; but even this is of more doubtful propriety than our English one of the *common chord*. Its being the more *common* or frequent is a matter of fact, on which it is better to rely than on its *perfection*, which is of doubtful meaning. Perhaps, to call it the *principal* (or *fundamental*) chord would be better still.

(R), p. 32. This fact was noticed by Mersenne towards the beginning of the 17th century, and afterwards more distinctly by Sauveur (Whewell, Induct. Sci. vol. ii, p. 394). It is to be observed, that though the 8th, 12th, and 17th are the sounds most prominently heard, yet the sounding body also yields many others. "Le corps sonore," says Rousseau (Dict. de Musique Harmonie), "ne donne pas seulement outre le son principal les sons qui composent avec lui l'accord parfait, mais une infinité d'autres sons formés par toutes les aliquotes du corps sonore, les quels n'entrent point dans cet accord parfait."

(S), p. 33. This chord is called by the German writers the *diminished triad*. But, independently of the objection reasonably attaching to the term *triad*, as applied to a chord of 3rd, 5th, and 8th (as to which, see note Q), it is both inconvenient and incorrect to apply the term *diminished* to any but such extraneous intervals as consist of intervals of the scale diminished by a chromatic semitone (v. sup. p. 28). The term *imperfect common chord* is free from these objections, and is fully warranted by the authority of various authors. This chord, when it occurs in certain connections, is sometimes considered as the chord of the dominant 7th *with the bass omitted*. But to suppose the bass omitted in any chord, seems contrary to the principles of harmony, as it is the bass from which the intervals of every chord are counted, and from which, therefore, it derives its individual character and denomination.

(T), p. 34. It is to the French theorist, Rameau (born in 1683), that the credit belongs of having introduced the classification by which some chords are considered as *direct*, and others as mere *inversions* of them, because consisting of the same notes, though the bass notes of the latter chords be different, so as to entitle them in that respect to the character of independent chords. This classification is of great practical use in the science of harmony—at least, as regards the particular chords mentioned in the text, viz. the common chords, major, minor, and imperfect, and the chords of the 7th, to which also is to be added the chord of the diminished 7th (as to which, v. Table, p. 69, No. 40). For all these it is very convenient to

P

consider as *direct* chords; and certain others, differing from them by the mere transposition of the bass notes, as their *inversions*. The classification, indeed, is often carried further, and applied to other chords besides these; but this is attended, in the opinion of the present writer, with no advantage.

(U), p. 34.   To Rameau we are also indebted for developing the idea of a *fundamental bass*, as distinguished from an actual bass; an idea, indeed, that immediately resulted from the fact already known in his time, that any note, when struck, produces not only its principal sound, but those also of its 8th, 12th, and 17th ascending; or, in other words, its common chord major (v. sup. p. 32).   For it was a corollary from this, that if any common chord major be taken, its actual bass (or lowest note) is also its generator, and the generator also of each of its notes individually considered—and, therefore, also of its several inversions; or, as it is otherwise expressed, the *fundamental bass* (or *root*) of all these; while, on the other hand, this fundamental bass is not the actual bass of any of the inversions; for in every inversion the actual bass of the direct chord is transposed; nor is it necessarily the actual bass of any of the notes individually considered.   To refer chords and notes in this manner to the same fundamental bass, though the actual bass be different, much facilitates the study and practice of harmony; and not only as regards common chords major and their inversions (to which alone the idea properly and strictly applies), but also as regards common chords minor and imperfect, and chords of the 7th and their inversions respectively, to which it may be conveniently made to apply for practical purposes.   But these are the only chords to which, in the author's judgment, a fundamental bass can be usefully assigned; for, in any other but these, the departure from the original idea is so great as to produce great uncertainty and embarrassment in its application.   Many writers, however, are not satisfied without attempting to find a fundamental bass for *every* chord without exception; while, on the other hand, with many others, the whole doctrine of fundamental bass is fallen into discredit or neglect—the natural effect of its being carried too far by those who consider it as applicable to every chord, and of its having also been formerly pushed to extravagant lengths by Rameau and his followers, who interweaved with it certain rules for the *progression* of the fundamental bass,—rules founded on visionary views and imposing restrictions more severe than those of nature herself.

(V), p. 36.   This, as far as the 7th or *leading note* is concerned, is contrary to the view taken by some writers, who allow no notes to be fundamental but those which have a common chord major or minor—a view that excludes the leading note, the common chord of which is imperfect.   " Every radical bass," says Callcott, " must have a perfect 5th."   To treat the leading note, however, as a fundamental bass, seems scarcely more of a licence than to treat the 2nd, 3rd, or 6th as such; and convenience and uniformity are much consulted by thus ranking *every* note as a fundamental bass to its own chord of 3rd, 5th, and 8th. This agrees in substance with the doctrine of a French writer, M. Rey. " Quoique la note qui sert au septième degré majeur et au deuxième du mineur ne soit pas reputée fondamentale parcequ'elle porte une quinte diminuée (meaning a minor 5th), et quoique les accords formés sur ces deux degrés ne soient regardés, que comme incomplets, ils font la fonction d'accords fondamentaux quand ils sont liés par une marche diatonique fondamentale de tierce ou de quinte. Sans cela, l'harmonie ne pourroit s'etablir dans toute l'étendue de la gamme." And still more distinct authority for the arrangement in the text is to be found in Heck's System of Harmony, who, in treating of the *imperfect common chord*, speaks of the 7th of the key (in the major mode) as its *fundamental bass*. See also, G. Weber, on Composition, translated by Warner, p. 167.

(W), p. 36. " The tonic, dominant, and subdominant, belong the more strictly and intimately to every key as its principal elements ; they are the heads of the family ; they determine its character, and they most distinctly impress the key on the ear," &c. (G. Weber, p. 260.) The same writer also designates the harmonies of the tonic, dominant, and subdominant, as the *essential*, and those of the other notes as the *accessory* harmonies of the key. Ibid, p. 259, 284.

(X), p. 37. The question as to the *origin* of the minor mode has always been treated as one of great difficulty. Yet it seems obvious that the constitution of this mode would be at once suggested by the simple conception of reversing the arrangement of the harmonies of the major mode (as to which, v. sup. p. 35) ; or, in other words, of attaching common chords minor, instead of major, to the primary fundamental basses of that mode. For example, while the key of C major may be formed, by taking common chords *major* upon its primary fundamental basses, C, F, G (thus C E G $\overline{\text{C}}$; F A C $\overline{\text{F}}$; G B D $\overline{\text{G}}$), the notes of which, placed in proper order, produce the scale in C major, viz. C, D, E, F, G, A, B, $\overline{\text{C}}$, the key of C minor may be formed by taking common chords *minor* on the same fundamental basses (thus, C, E♭, G, $\overline{\text{C}}$; F, A♭, C, F; G, B♭, D, $\overline{\text{G}}$); the notes of which, placed in proper order, produce C, B♭, A♭, G, F, E♭, D, C, the scale of C minor descending. This sufficiently accounts for the scale of the minor mode in its descending progression; and as to the augmentation of the 6th and 7th by a chromatic semitone, in the ascending progression, an explanation has been offered of this in a former place (v. sup. note G). We shall also have occasion, in a later part of the work, to point out a distinction of some curiosity between the two modes, as regards the *ratios of vibrations*, on which the arrangements of their intervals are respectively founded (v. p. 185).

(Y), p. 38. It deserves notice, that 3rds and 6ths are concords capable of alteration ; viz. by the difference of a semitone (as expressed by the adjuncts of *major* and *minor*), without ceasing to be concords; but that octaves, 5ths, and 4ths, admit no such alteration, without becoming discords. This distinction is usually expressed by calling the former *imperfect* and the latter *perfect* concords; but it would be more correct, perhaps, to employ the terms *variable* and *invariable.*

As to the 4th, it is to be observed that its quality has been the subject of much controversy, there being many writers who hold it a discord. But the classification of it as a concord is much more common, and is warranted by the better authorities. According to the definition given in the text of concordant intervals—viz. that they are such as are comprised within some common chord major—the 4th clearly belongs to that class; for there is always some common chord major in which every 4th is comprised. Thus, the 4th, G C̄, is comprised within the common chord major, C, E, *G,* C̄. But it is to be observed that the place which the 4th occupies in the common chord is *between two upper parts,* as in the example just given, and not *between the bass and an upper part;* and that when the 4th occurs, in music, in a position different from that which it occupies in the common chord—viz. when it occurs between the bass and an upper part (as in the chord of $\frac{6}{4}$, *G, C,* E, C̄)—it strikes the ear as a sort of discord, and apparently for this reason, that it is felt as a concord *out of its right place.* This subject will be resumed hereafter (v. post, note WW).

(Z), p. 40. "These discords, the tritone and semi-diapente" (the old name for the minor 5th), "as also the 2nds and 7ths, are of very great use in music, and add a wonderful ornament and pleasure to it, if they be judiciously managed. Without them, music would be much less grateful, like as meat would be to the palate without salt or sauce." (Holder, on Harmony, p. 169.) With respect to 2nds, however, it is to be observed, that the term 2nd is to be understood here in the sense in which it is always used in the present treatise; viz. as applying to a note which differs from another by the interval of a *whole* tone; for though a note which differs from another by the interval of a *semitone* only, is in some sense a 2nd to that note (being removed from it only by a single degree), yet it is a 2nd of a very different quality, and unfit, generally speaking, for harmonical use. "The less second, or semitone," says another writer, "is a much more harsh and impracticable discord than the greater 2nd, or whole tone. The former of these can only be admitted seldom, and with great caution; but the latter is very often introduced. Thus the 7th and key-note on the 3rd and 4th of the scale are seldom joined in harmony, these being intervals of a semitone only; but the key-note and 2nd, or the 4th and 5th, or the 5th and 6th, or the 6th and 7th of the scale, are often set together, these being all intervals of a whole tone." Holder, on Music, p. 55.

(AA), p. 45. These intervals, as stated in a former place (sup. note N), are usually described, not as "extraneous" intervals (the term which has been adopted in this work, as the most significant and appropriate, because it expresses their departure from the proper scale), but as "altered" intervals. And so the discords now in question, to which the denomination of "extraneous" is also assigned in this work, are commonly described as "discords with altered intervals," or as "discords with augmented or diminished intervals." They can scarcely be said, however, to have any fixed denomination, owing to the imperfect notice hitherto taken of them in the treatises.

(BB), p. 49. The extraneous discords contribute not only to the variety of music, but very remarkably also, in some cases, to its point and energy. (See the remarks on the extreme sharp 6th and diminished 7th, post pp. 160, 170.) They constitute, however, in general, a *harsher* class of discords than that of discords of the scale, as will more fully appear hereafter; v. pp. 189, 195.

(CC), p. 50. According to some writers, the bass is no exception; for they speak of certain combinations as being certain chords *with the bass omitted*. But as a chord is characterized only by its intervals, counted from the bass, the idea of a chord without its bass seems to be inadmissible. It is easy to understand that B, D, F, is the imperfect common chord, B, D, F, $\overline{\text{B}}$, with the 8th, B, omitted; but to consider the chord of the 7th on the leading note, B, D, F, A, as the chord of $\begin{smallmatrix} 9 \\ 7 \\ 5 \\ 3 \end{smallmatrix}$ G, B, D, F, A, with the bass, G, omitted, according to the method of some treatises, is contrary to principle, for the reason just assigned.

(DD), p. 71. According to A. Reicha, the following eight concords and discords are those most frequently and regularly used within the compass of the same key, in the major mode:

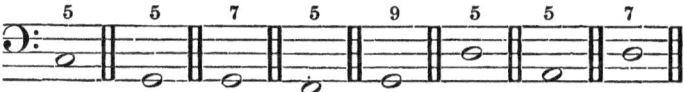

But to these the inversions of each are to be added.

(EE), p. 71. This will be apparent on examination of the arrangement of the common chords of the key, sup. p. 35, where it will be found that the columns whose fundamental basses are a 2nd or 7th from each other have no note in common; that those whose fundamental basses are at the interval of a 4th or 5th, have one; and those whose fundamental basses are at the interval of a 3rd or 6th, have two.

(FF), p 72. A formula has long been in use among musicians, called, in France, where it was invented, *la règle de l'octave*; in Italy, *la scala armonica*; in England, *the natural harmony of the scale*. It professes to give a rule for accompaniment, not to a given *melody*, but to a given *bass* ; viz. to the scale itself, taken as a bass ; and an acquaintance with it has generally been considered as very useful to the student in harmony ; though, on the other hand, there are parts of its progression which have not escaped criticism. (See Rousseau's Dict. de Musique, Art. Règle de l'Octave ; Callcott's Grammar, p. 243 ; and Fétis's Harmonie, p. 86.

This formula, in the shape in which it is most usually given, is as follows :

Règle de l' Octave, in the Major Mode.

Shield.

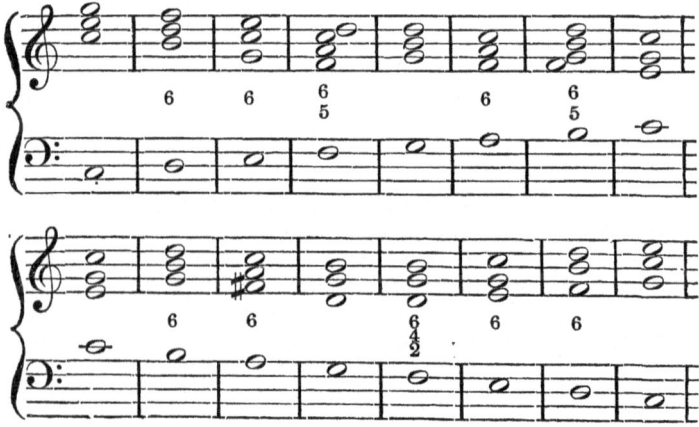

It seems to have been first published by M. Delaire, in the year 1700, when its shape is said to have been as follows, which it may be interesting to compare with the modern one :

(GG), p. 86. The minor mode (often called, though improperly, the minor *key)* is universally felt to be expressive of melancholy or pathetic sentiment, and to be distinguished in this respect from the major. This character is the more strongly marked when the minor mode is introduced, by way of change or modulation from the major, so as to be contrasted with it, which may take place either by a change from a major key into the *relative* minor key, or into the minor key of the same tonic, or into that of a different tonic.

(HH), p. 88. It is to be observed that the rule speaks of 5ths—that is, true 5ths, or 5ths properly so called. The succession of a minor 5th upon a true one, by similar motion, is not uniformly prohibited. "An imperfect 5th," says Shield (meaning a minor fifth), "is allowed to follow a perfect in descending, provided the highest note falls and the lowest rises afterwards; thus:

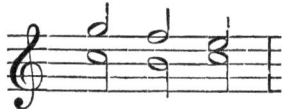

(II), p. 92. Of all the prohibitory rules in music, that which relates to consecutive 5ths and 8ths has ever been the most prominent and the most universally received. It is therefore the more important to investigate the principle on which it rests. But that principle has never yet been fully agreed upon. Indeed, a recent writer of considerable note represents it as involved in total obscurity. "On ne sait pas encore," says A. Reicha (Harmonie, p. 132), "pourquoi des quintes, de suite par movement semblable, produisent un mauvais effet." Various reasons for this rule, however, have been assigned by various authors. Kircher, in his Musurgia, published in 1650, considers consecutive intervals of the same kind as objectionable, simply on the ground of their *want of variety (quod nullam varietatem habeant) ;* and this explanation of the matter, which is perhaps the earliest, seems also to be the best that has been offered. It is observable that at one period consecutive *thirds* (now in every case freely admitted) were made the subject of a similar prohibition (Rousseau's Dict. de Musique, Art. Tierce); and the case was the same (as, indeed, it still is) with consecutive *fourths,* unless combined with sixths and thirds ; and no reason has ever been suggested for the rule that in its nature would be applicable to *all* these cases, except the want of variety, which, on the other hand, *does* apply in a greater or less degree to them all. Afterwards, when the restriction began to relax with respect to thirds, the course of the relaxation indicates the same principle. Thus the succession of a major and a minor 3rd exhibits somewhat more variety than that of two major 3rds ; and accordingly, in Holder's time, who wrote in 1694, it was already held that the former was entirely unobjectionable, but the latter was to be used sparingly, and, like a discord, only "to give, after a little grating, a better relish" (Holder, Treatise on Harmony, p. 85). The want of variety seems also to be pointed at by the present allowance, even in the case of 5ths, of such as follow *by contrary motion*—an allowance for which the most satisfactory reason that can be suggested seems to be that a contrary motion has more variety than a similar one. The same view is also confirmed by the exception above referred to in the case of consecutive 4ths; viz. that of their becoming allowable when combined with 6ths and 3rds; for as 6ths and 3rds are variable intervals, that is, capable of being taken sometimes as major and sometimes as minor, their intermixture in these different forms suffices to cure the monotony which would result from the 4ths if taken alone. Rousseau observes (Dict. Art. Quinte) that Rameau endeavours to account for the rule, on the ground of want of *liaison;* but says—" Il se trompe; premièrement, on peut former ces deux quintes et conserver la liaison harmonique; secondement, avec cette liaison les deux quintes sont encore mauvaises; troisiemement, il faudroit par le même principe étendre comme autrefois, la règle, aux tierces majeures; ce que n'est pas, et ne doit pas être."

(KK), p. 92. Though this rule is both of an important and a fundamental character, it is very little noticed by our English writers. In the Treatise, indeed, of the Rev. W. Jones, of Nayland, p. 59, *liaison* is substantially referred to, but rather in the way of recommendation than of rule. He says—" It is a general rule, that an agreeable succession is produced when the preceding and succeeding chords are connected by some note common to both." The French writers are more distinct on this subject. In laying down, as they do, the rule which prescribes generally a progression by consonant intervals, they assign, as the reason, that the chords would otherwise have no *liaison*. That, in fact, they would not, we have already shown to be the case (v. sup. p. 71, note EE).

(LL), p. 94. It is by an attentive observation of examples that the manner of the resolution of discords is best collected, and it is difficult to lay down rules which will not be subject to an embarrassing number of exceptions. In the present work will be found, from p. 135 to p. 178, examples of all the principal discords, and of the manner of their resolution in the particular connections in which they there occur. The attentive study of these, with the aid of the general explanations afforded at pp. 94, 95, 96, will give a considerable insight into the subject of resolution, and such as is not capable of being improved by any attempt to lay down particular rules.

(MM), p. 97. The reason assigned for this rule by Burrowes, is that the doubling tends to produce consecutive octaves, because each of the doubled notes will descend a semitone in the resolution that follows, and therefore there will be octaves again. But the more obvious reason assigned in the text seems sufficient. The rule is subject to several exceptions, among which is the case of the chord of the $\frac{5}{2}$; where the 2nd may be doubled, as in the following example:

Albrechtsberger.

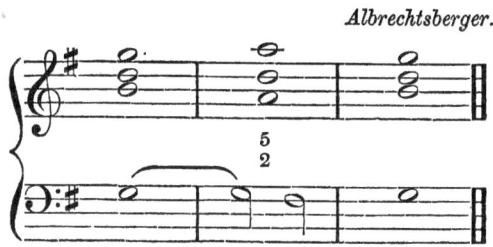

(NN), p. 98. The first inversion of the chord of the 7th on the 2nd (the second chord in the example in the text), and which is there resolved by the descent of the 5th while the 6th remains, is also capable of being resolved in another way; viz. by the ascent of the 6th while the 5th remains, making a cadence on the tonic. Thus:

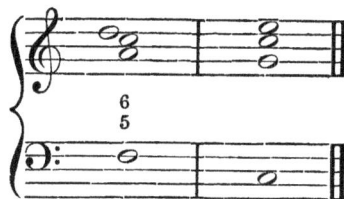

In this latter case, Rameau and his followers made it a different chord from the first inversion of the chord of the 7th on the 2nd (though the two chords were no otherwise distinguishable, except by the manner of the resolution), and called it the *chord of the added 6th*, considering it as the common chord of the subdominant *with the 6th added*. This anomalous arrangement, adopted only to justify some fanciful doctrines of Rameau with respect to the progression of the fundamental bass, has been generally followed also in the English treatises. It is time, however, to lay it aside, and to classify this chord, in every case, and whatever may be the mode of its resolution, as the first inversion of the chord of the 7th on the 2nd. It is true that the inversion is, in this instance (as in some others), of more importance, and in more frequent use, than the direct chord itself: but this affords no argument against the classification.

(OO), p. 100. The following examples of the full cadence are from a French treatise of considerable note:

(PP), p. 101. The words *authentic* and *plagal* are derived from the old ecclesiastical music. They occur, at present, rather in treatises than in familiar musical language. The plagal cadence, or cadence from the sub-dominant, is used principally in Church Music. Callcott refers to examples of it in the Hallelujah Chorus and Coronation Anthem.

(QQ), p. 101. The *imperfect* cadence seems to have been formerly called a *middle* cadence. For Pepusch (the contemporary of Handel), who has been described as the most *orthodox* writer of his day, thus expresses himself: " Cadences in music are the same as stops in speaking or wrtiing; for which reasons they are distinguished into full cadences and middle ca-dences. These last are like commas and semicolons, after which more is expected to follow, they not making so full a stop as the others; whereas, after a full cadence, we are sensible that we are come to a conclusion."

(RR), p. 104. Modulation is one of the chief sources of musical ex-pression; and there is one particular course of modulation which is some-times so powerfully applied to purposes of pathos, that it seems to deserve specific notice; viz. that which consists in introducing a note flatter by a semitone than the antecedent progression in the original key would naturally lead the ear to expect; as in the following examples—the first, remarkable for the thrilling sadness of its B flat; the second, for the religious awe and veneration of its C natural :

1st **Example.**—A modulation from the key of C major into the key of F major:

2nd **Example.**—A modulation from the key of D major into the key of G major:

It is to be observed, however, that an unexpected semitonic descent of this kind is also capable of being produced *without* a modulation; viz. by the mere employment of an extraneous discord introductory of no new key; and that when so produced, its power is as perceptible as in the former case. In the following example, so finely expressive of mournful indignation, the effect is due to the unexpected introduction of C natural in the chord of the diminished 7th, D♯ F A C, which takes place without any modulation:

(Elijah) Mendelssohn.

el have broken thy covenant, broken thy covenant

(SS), p. 105. It may be convenient to exhibit here, by way of example, the keys that are most nearly related to the key of C major and A minor respectively, with their order of propinquity :

C major {
A minor (or key of the 6th note),
G major (or key of the dominant),
F major (or key of the subdominant),
E minor (or key of the 3rd note),
D minor (or key of the 2nd note).

A minor {
C major (or key of the 6th descending),
E minor (or key of the dominant),
D minor (or key of the subdominant),
F major (or key of the 3rd descending),
G major (or key of the 2nd descending).

The order of propinquity, however, is, in some respects, differently stated by different writers.

(TT), p. 130. In the Treatise of the Rev. W. Jones, of Nayland, he expresses an opinion that though the position of C E G $\overline{\text{C}}$ is the most *natural*, as affording an accompaniment to the key note, yet that the most *harmonious* position is C G $\overline{\text{C}}$ $\overline{\text{E}}$; for he argues that, instead of E G the minor 3rd, and E $\overline{\text{C}}$ the minor sixth, comprised in the former arrangement, the latter presents us with G $\overline{\text{E}}$ a major 6th, and $\overline{\text{C}}$ $\overline{\text{E}}$ a major 3rd; and instead of C E a *third* from the key note, C $\overline{\text{E}}$ a *tenth* from the key note, " more agreeable to all ears than a 3rd."

(UU), p. 133.   This chord of the 6th upon the subdominant, when employed preparatory to a cadence, has been classed by some writers as a common chord on the subdominant, with a 6th substituted for a 5th, and consequently called the *chord of the substituted 6th*.   This denomination rests on the same principle as that of the *added 6th*, to which we have had occasion already to refer (v. sup. note NN).   And it is equally opposed to sound classification.   This chord ought, in every case, to be considered as the first inversion of the common chord minor upon the 2nd of the key.

(VV), p. 133.   The following example of chords of the 6th will show that a series of them is capable of being played in immediate succession, and even by similar motion, with very pleasing effect: a property of these chords that well deserves remark :

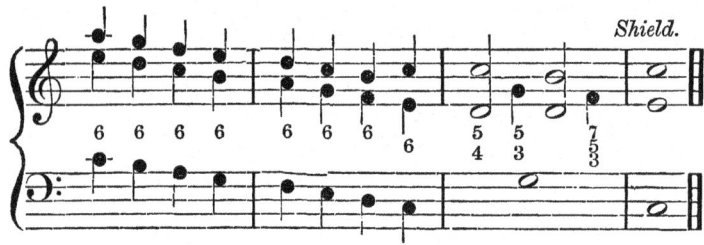

Shield.

This effect is due to the quality of each of those intervals of 6th and 3rd of which a chord of the 6th consists.   For both a 6th and a 3rd are capable of being played consecutively in the various forms of major and minor, as in the above example; and therefore a succession of either or both may take place without incurring the objection to which a series of the invariable intervals of perfect 5ths or 8ths is liable, on the ground of want of variety ; while, at the same time, a major and a minor 6th (or a major and a minor 3rd, as the case may be), being both intervals of the same number of degrees and both concords, have a close family resemblance to each other, such as make a group of them interesting to the ear.   For we shall constantly find it to be true, that the combination of order or symmetry, with variety, is the great principle on which the pleasure of music depends.

As the pleasing effect of a succession of chords of the 6th arises from the quality of each of the intervals of 6th and 3rd, and not merely from their combination, so we shall find that a succession of either of these intervals is not less agreeable, though they may happen to occur as constituent parts of *other* chords, not being chords of the 6th, as in the following example, where a series of 6ths occur in this manner between the singing parts :

Un poco Andante. (Magdalena) G. Hasse.

Cle-men-ti-a tu-a in-fi-ni-ta par - ce do-

Cle-men-ti-a tu-a in-fi-ni-ta par - ce do-

len-ti, O De - us! par - ce do-len-ti, O De-us!

len-ti, O De - us! par - ce do-len-ti, O De-us!

(WW), p. 133.   This rule is strongly insisted on by A. Reicha, who remarks, with truth, that it is generally left unnoticed in the treatises (see his Traité d'Harmonie, p. 16).   As to the reason, however, on which the rule is founded, he is silent; and the following remarks on the subject may therefore be acceptable.   The necessity for *preparing* the 4th, when its lower note is the bass of the chord, must clearly result from some harshness with which that interval is found to be attended when so circumstanced; for *preparation* is an expedient adopted to reconcile discord to the ear.   That the 4th, how-ever, should in any case stand in need of such reconcilement, is a considera-tion attended with some difficulty, because it forms part of the common chord, and is, by virtue of that circumstance, a concord (v. sup. p. 38, and Note Y).   But the reason is, that though it occurs in a common chord, it occurs there only between two *upper parts;* and therefore, when it is found *between an upper part and the bass,* it has a collocation foreign to the common chord, or, in other words, it is a concord *out of its right place.*   It may be objected, perhaps, that, according to this view, the *minor 3rd,* when its lower note is the bass of the chord, would also require *preparation;* for when we take a chord of the 6th—for example, *E G* C G—the minor 3rd, E G, formed between the bass and one of the upper parts, is out of its proper collocation in the common chord, *C E G* $\overline{\text{C}}$, where it occurs between two upper parts; and yet no preparation has ever been deemed necessary in the case of a minor 3rd. But the anwer is, that the analogy between the two cases is not complete. Neither the common chord major nor the common chord minor has the interval of a 4th between the bass and an upper part; but the common chord minor has a minor 3rd so placed; for example, E G occurs between the bass and an upper part in the common chord minor, *E G* B E.   This explanation of the prepared 4th is believed to be quite new; the reader will judge for himself how far it is satisfactory.   Supposing it to be so, it sets at rest the much controverted question whether the 4th is a concord or discord.   It must be ranked as a concord generally, because part of the common chord; but it is, nevertheless, of a somewhat dissonant character when out of its place in that chord, and may be said to be, in that case, a *quasi* discord.

(XX), p. 142.   As to the first inversion of the chord of the 7th on the 2nd of the key, vide sup. note NN.   This and the first inversion of the chord of the dominant 7th are the only chords of $\frac{6}{5}$ which are in frequent use.   The following example, however, includes some of the others; viz. the first inversion of the chord of the 7th on the tonic (or $\frac{6}{5}$ on the 3rd of the key), and the first inversion of the chord of the 7th on the leading note (or $\frac{6}{5}$ on the 2nd of the key):

Albrechtsberger.

(YY), p. 159. A glance at the Synopsis, pp. 15 and 16, will show that the intervals of augmented major 6th and diminished minor 6th, and therefore of course the intervals of which these are respectively the inversions, viz. the diminished minor 3rd and the augmented major 3rd, are all strictly extraneous, that is, are not to be found in either mode, whether in the ascending or descending series. From this it follows also that all chords, comprising any of these intervals, are in like manner strictly extraneous. Among these are included the chord of the extreme sharp 6th, or of the extreme sharp 6th with a 5th or with a tritone (Table, p. 67, Nos. 36, 37, 38), the chord of the minor 7th, minor 5th, and major 3rd (Table, p. 65, No. 35), the chord of the diminished minor 7th, diminished 5th, and diminished minor 3rd (Table, p. 69, No. 44), and the chord of the diminished minor 7th, diminished minor 6th, and minor 3rd (Ibid. No. 45). But the last three chords are of extremely rare occurrence.

(ZZ), p. 160. According to Rousseau, this chord " ne se pratique jamais que sur la sixième note d'un ton mineur descendant sur la dominante." But it is not to be inferred from this that the chord belongs to the minor mode, which, for the reason given in the text (p. 159), cannot correctly be said. Neither are we to understand, by the expression *sixième note*, the 6th note ascending, and bearing (as it does in the entire series of the scale) a chromatic augmentation. For on such a 6th this chord could not be taken. It is to the 6th ascending, and without augmentation, that the proposition must be intended to refer— for of no other is it true.

(AAA), p. 165. That the consecutive 5ths, in this instance, are entitled to toleration at least, seems to be generally admitted by musical ears. Various reasons have been assigned for this peculiarity; among others, that the consecution is, in this case, occasioned by the resolution of a discord (Gunn's Introduction to Music). But why the progression should be rendered inoffensive by that circumstance, it is difficult to perceive. The true ground on which the ear tolerates the successive 5th in this case, may perhaps be discovered by referring to the principle on which the rule against successive 5ths is itself accounted for in a former note (see note II). It is the *want of variety* involved in a repetition of the same interval between the same parts that was there supposed to be the real objection felt by the ear to such a succession in general. But in the particular succession now in question, though each chord contains a 5th between the same parts, yet there is so far a variety, that the first chord is *extraneous*, and the second *belongs to the scale;* and that even a slight variety affords a dispensation from the rule, is sufficiently proved by the case where 5ths succeed by way of *contrary motion*, which, as before explained (p. 91), is received as a legitimate progression.

(BBB), p. 168. We may remark here, that, of the three chords of the extreme sharp 6th that have been mentioned, the first is sometimes called the Italian 6th, the second the German 6th, the third the French 6th. See Callcott's Grammar, p. 238 ; who adds, that the music of Italy, Germany, and

Q

France cannot be illustrated in a smaller compass than by the use of these chords. In Gunn's Introduction to Music, p. 224, it is said that though the third of them is generally prescribed and exemplified by masters of the French schools, it very seldom appears in musical composition.

(CCC), p. 179. In reference to the subject of this chapter, it will be convenient to explain the method of calculating ratios; for which we may refer generally to Holder on Harmony, and Malcolm on Music.

1. *To add one ratio to another.*—Turn them into fractions, and multiply continually the numerators into the numerators, and the denominators into the denominators. The last product is the answer. Ex. Add 2 : 3, 5 : 7, and 8 : 9 together, $\frac{2 \times 5 \times 8}{3 \times 7 \times 9} = \frac{80}{189}$ or 80 : 189 the answer.

2. *To subtract one ratio from another.*—Turn them into fractions, and multiply them crosswise. Ex. Subtract 3 : 2 from 5 : 3 .. $\frac{5}{3} \times \frac{3}{2} = \frac{10}{9}$ or 10 : 9 the answer.

3. *To multiply a ratio by a whole number.*—Add the ratio to itself (by the rule) as often as is expressed by the whole number, and the sum will be the answer. Ex. Multiply 2 : 3 by 4 .. $\frac{2}{3} \times \frac{2}{3} \times \frac{2}{3} \times \frac{2}{3} = \frac{16}{81}$ or 16 : 81 the answer.

4. *To divide a ratio by a whole number.*—Find such a ratio as, being added to itself as often as the whole number expresses, will produce the given ratio, and this will be the answer. Ex. Divide 9 : 16 by 2. $\frac{9}{16} = \frac{3}{4} \times \frac{3}{4}$. Therefore $\frac{3}{4}$ or 3 : 4 is the answer.

(DDD), p. 179. See Holder, on Harmony; Stillingfleet, on Harmony; Rousseau's Dict. de Musique, Art. Echelle; Arnott's Physics, vol. i, p. 511; and other authorities. If these fractions be reduced to a common denominator of 24, the ratios may be expressed as follows:

C	D	E	F	G	A	B	$\overline{\text{C}}$
24	27	30	32	36	40	45	48

(EEE), p. 180. This may be proved from the ratios to the key note (sup. p. 179), as follows:

$$C = 1; \quad D = \frac{9}{8}; \quad E = \frac{5}{4} = \frac{9}{8} \times \frac{10}{9}; \quad F = \frac{4}{3} = \frac{5}{4} \times \frac{16}{15};$$

$$G = \frac{3}{2} = \frac{4}{3} \times \frac{9}{8}; \quad A = \frac{5}{3} = \frac{3}{2} \times \frac{10}{9}; \quad B = \frac{15}{8} = \frac{5}{3} \times \frac{9}{8};$$

$$\overline{C} = 2 = \frac{15}{8} \times \frac{16}{15}$$

(FFF), p. 180. Thus:

$$A = 2; \quad G = \frac{9}{5} = 2 \times \frac{9}{10}; \quad F = \frac{8}{5} = \frac{9}{5} \times \frac{8}{9}; \quad E = \frac{3}{2} = \frac{8}{5} \times \frac{15}{16};$$

$$D = \frac{4}{3} = \frac{3}{2} \times \frac{8}{9}; \quad C = \frac{6}{5} = \frac{4}{3} \times \frac{9}{10}; \quad B = \frac{9}{8} = \frac{6}{5} \times \frac{15}{16}; \quad A = 1.$$

(GGG), p. 180. Thus:

$$\overline{A} = 1; \quad B = \frac{9}{8}; \quad C = \frac{6}{5} = \frac{9}{8} \times \frac{16}{15}; \quad D = \frac{4}{3} = \frac{6}{5} \times \frac{10}{9};$$

$$E = \frac{3}{2} = \frac{4}{3} \times \frac{9}{8}; \quad F\sharp = \frac{5}{3} = \frac{3}{2} \times \frac{10}{9}; \quad G\sharp = \frac{15}{8} = \frac{5}{3} \times \frac{9}{8};$$

$$\overline{A} = 2 = \frac{15}{8} \times \frac{16}{15}.$$

(HHH), p. 181. These ratios between contiguous notes in the minor mode, are shown on the face of the calculations—sup. notes (FFF), (GGG). One point worthy of notice, that the examination of these ratios suggests, is that, in passing from any key in the major mode to its relative minor, or *vice versa*, an alteration takes place in some of the intervals. Thus, in the key of C major (see p. 180), the interval from C to D is a major tone, and the interval from D to E a minor tone; but in the relative key of A minor (see p. 181), the interval from C to D is a minor tone, and that from D to E a major.

(III), p. 181. "Most people," says Stillingfleet, in his Treatise on Harmony, p. 32, "will be apt to think that there was not much thought required to settle the common octave, which almost any one that has an ear can run over with the greatest ease, and, as he thinks, naturally. Yet there were many divisions of it proposed before that was invented which now takes place. Ptolemy, the astronomer, was the inventor; and it is no wonder that it has generally prevailed from his time to this day, as it is the only one which was truly founded in nature. We are taught to go through it after a manner, and are even often apt to look upon it as natural; but it is undoubtedly artificial, and the result of much profound thought. However paradoxical, therefore, it may seem, it is certainly true that harmony is more natural than the notes of the octave; for a string cannot be sounded without producing harmony." (He alludes here to the natural generation of the common chord, as explained sup. p. 32.) " Whereas the notes of an octave never appear but in highly civilized countries. Amongst the birds, we hear the 5th, the 4th, the 3rd major and minor; but the notes of the octave from no animal that has not been taught," &c.

The fact stated in the course of this extract, that many different scales preceded that which now prevails in civilized Europe, is abundantly proved by the Greek treatises on music, still extant, and collected by Meibomius; and by the accounts given even of the modern music of many of the oriental nations. Thus, in the preface to the learned Treatise of M. Fétis, it is said that among the Persians the scale was, from an ancient period, divided into quarter tones; and that the Persian musicians whom Amurath IV carried captive to Constantinople, after the taking of Bagdad, in 1638, introduced this scale into Turkey, where it was still in use at the close of the 18th century; and the Letteratura Turchescha of Toderini is cited in support of this statement. That the present scale, therefore, was in the nature of a progressive invention, and not a system of sounds that nature instinctively

and at once inculcates, is incontestible. The claim of Ptolemy, however, to
be considered as the inventor is questionable. The honour seems rather to
belong to Didymus, of Alexandria, who preceded him; and who, as appears
by the writings of Ptolemy himself, had already suggested the interval of
the minor tone, 10:9, and of the diatonic semitone, 16:15. For it was
the suggestion of these intervals that was the great desideratum; the major
tone of 9:8 having formed part of the system of the more ancient Greeks,
though it would seem to have comprised neither of the others. These once
introduced, and all the elements of the scale being thus complete, the rest
of it was evidently mere matter of arrangement.

(JJJ), p. 186. It is to be observed that one of the 3rds in the scale,
viz. that betweeen the 2nd and the 4th (for example, D F, in the key of C
major), though nearly approaching to a minor third of $\frac{6}{5}$, falls short of it,
and has the ratio of $\frac{32}{27}$ only, which is less than $\frac{6}{5}$ by a comma.   For
$D E \times E F = \frac{10}{9} \times \frac{16}{15} = \frac{32}{27}$; and $\frac{6}{5} - \frac{32}{27} = \frac{81}{80}$. This may be
considered, therefore, as an imperfect or defective minor 3rd;—and it is a dis-
cord;—in which respect it again differs from the true minor 3rd, which is a
concord.

(KKK), p. 187. This, according to the best authorities, is the ratio of
the major 6th. It is stated, however, by some writers as $\frac{27}{16}$; the scale, according to them, being

C	D	E	F	G	A	B	$\bar{\text{C}}$
1	$\frac{9}{8}$	$\frac{5}{4}$	$\frac{4}{3}$	$\frac{3}{2}$	$\frac{27}{16}$	$\frac{15}{8}$	2

which makes the ratios between contiguous notes—

C to D	D to E	E to F	F to G	G to A	A to B	B to C
$\frac{9}{8}$	$\frac{10}{9}$	$\frac{16}{15}$	$\frac{9}{8}$	$\frac{9}{8}$	$\frac{10}{9}$	$\frac{16}{15}$

And it is observable that this ratio of $\frac{27}{16}$ is the one which would, in fact,
belong to the 6th, if, in the formation of the scale, the multiplication by 3 were
carried one step farther than was supposed at p. 183; that is, if, after obtain-
ing the note D, by the multiplication of 3×3, the next step were to multiply
again by 3; for $3 \times 3 \times 3 = 27$, and 27 brought down so as to fall within
the compass of the first octave, would become $\frac{27}{16}$. It is also observable, as to
the arrangement of contiguous notes thus produced, that it would consist of
two disjunct tetrachords exactly similar; for the intervals in each case would be
$\frac{9}{8}$, $\frac{10}{9}$, $\frac{16}{15}$; but if we take $\frac{5}{3}$ as the ratio of the major 6th, the tetrachords
are *not* exactly similar, the intervals of the first being $\frac{9}{8}$, $\frac{10}{9}$, $\frac{16}{15}$; and of the

2nd, $\frac{10}{9}$, $\frac{9}{8}$, $\frac{16}{15}$ (v. sup. p. 180). On the other hand, however, it is to be remarked that $\frac{27}{16}$ is a *discordant* interval, and $\frac{5}{3}$ a *concordant* one. This circumstance alone seems sufficient to show that the latter is the preferable ratio.

(LLL), p. 187. The minor 7th is here taken as the interval between the 5th of the scale and the upper octave of the 4th in the major mode; for example, as G F in the key of C major. In this case, its ratio is $\frac{16}{9}$. For the ratios as between contiguous notes are as follows:

$$\left. \frac{\text{G to A}}{\frac{10}{9}} \right| \frac{\text{A to B}}{\frac{9}{8}} \left| \frac{\text{B to C}}{\frac{16}{15}} \right| \frac{\text{C to D}}{\frac{9}{8}} \left| \frac{\text{D to E}}{\frac{10}{9}} \right| \frac{\text{E to F}}{\frac{16}{15}} \right| = \frac{16}{9}$$

But if taken as the interval between the 3rd and the upper octave of the 2nd in the major mode—for example, as E D in the key of C major—its ratio is not $\frac{16}{9}$, but $\frac{9}{5}$. For the ratios as between contiguous notes then are :

$$\left. \frac{\text{E to F}}{\frac{16}{15}} \right| \frac{\text{F to G}}{\frac{9}{8}} \left| \frac{\text{G to A}}{\frac{10}{9}} \right| \frac{\text{A to B}}{\frac{9}{8}} \left| \frac{\text{B to C}}{\frac{16}{15}} \right| \frac{\text{C to D}}{\frac{9}{8}} \right| = \frac{9}{5}$$

(MMM), p. 188. Taking C as 1, C E is a major 3rd, $\frac{5}{4}$; and E G♯ is another major 3rd, $\frac{5}{4}$; and consequently, C G♯ is $\frac{5}{4} \times \frac{5}{4} = \frac{25}{16}$, the difference between which and $\frac{3}{2}$, the ratio of the 5th, C G, is $\frac{25}{24}$. But the difference between C G and C G♯ is the interval of the chromatic semitone G G♯. Therefore $\frac{25}{24}$ is the ratio of the chromatic semitone. It may be observed that this is less than the diatonic semitone of $\frac{16}{15}$ by $\frac{128}{125}$, being the ratio of the enharmonic interval.

(NNN), p. 189. If the minor 7th be taken as $\frac{16}{9}$ (v. note LLL), then the diminished minor 7th (which is the minor 7th diminished by a chromatic semitone) $= \frac{16}{9} - \frac{25}{24} = \frac{128}{75}$. But if the minor 7th be taken as $\frac{9}{5}$, then the diminished minor 7th $= \frac{9}{5} - \frac{25}{24} = \frac{216}{125}$.

(OOO), p. 191. The celebrated Tartini, in his Trattato di Musica, as cited by Stillingfleet, says : That when he plays double stops on the violin, " he can hit upon the very form itself of the intervals;" and has " the advantage for himself and his scholars of a sure intonation, and consequently of the real use of the scale, with all the precision of the true ratios."

(PPP), p. 192. If four successive 5ths ascending be taken, they will produce intolerably sharp major 3rds, which may be proved thus :

The ratio of the note which would be produced by the first 5th, or 5th from the pitch note, being $\frac{3}{2}$, the ratio of the note which would be produced by four successive 5ths would be

$$\frac{3}{2} \times \frac{3}{2} \times \frac{3}{2} \times \frac{3}{2} = \frac{81}{16};$$

which note would form the 5th to the key note in the third octave ascending. And it would approximate to the note forming the major 3rd to the key note in the same octave, but would be sharper by a comma, that is, by 81 : 80 (v. sup. p. 190).   For the ratio of the major 3rd from the pitch note being $\frac{5}{4}$, the ratio of the note which would form the major 3rd to the key note in the third octave ascending, would be

$$\frac{5}{4} \times 2 \times 2 = \frac{20}{4} = 5 \text{ ; and } \frac{81}{16} - 5 = \frac{81}{80}.$$

Now a comma is a much wider departure from the true interval than the ear is found, in the case of any concord, to endure.

(QQQ), p. 192. It is remarkable, that if we throw temperament aside and resort to an instrument exactly tuned to the true ratios of the scale, we not only encounter altered intervals (as explained at p. 190) in passing from one key to another, but we encounter also, within the compass of the same key, intervals which approximate to certain concords, but yet deviate from them by a comma, and which can therefore only be characterized as defective concords. Thus, between the 2nd and the 4th, we have a minor 3rd (as explained in note JJJ) defective by a comma.   And so in the interval between the 2nd and the 6th (for example, D A, in the key of C major), we have a 5th similarly defective.   For $D E \times E F \times F G \times G A =$ $\frac{10}{9} \times \frac{16}{15} \times \frac{9}{8} \times \frac{10}{9} = \frac{40}{27}$, which falls short of $\frac{3}{2}$, the ratio of a true 5th, by $\frac{81}{80}$. Whether we confine ourselves to a single key, therefore, or pass from one to another, music in exact tune is unattainable. But by temperament, both the inequalities in the same key, and those arising between different keys, are reduced; and a compromise effected, which, though it leaves scarcely any interval in the scale exactly true, yet is on the whole productive of a better intonation than can in the nature of things be otherwise attained.

(RRR), p. 196. That the pleasure produced by music, considered in general, consists in the perception of the order and proportion by which it is pervaded, is no new theory, having been maintained by the celebrated Euler, and by other writers; but in no other work, perhaps, than the present, has it been distinctly referred to the perception of order, proportion, and variety combined. Yet the more we examine the phenomena of musical effect, the more reason we still find for believing that this is the great source of musical pleasure; and though the charm of *particular progressions* depend, no doubt, in a great degree on causes of other kinds, to which some reference has already been made (v. sup. p. 193), yet, even by these, the truth of the same general principle is often strikingly confirmed. (See note VV.) It is illustrated too, as remarked in the text, by the analogous effect which the union of order and proportion with variety produces on the eye, in the contemplation of the works of art.

(SSS), p. 198. This theory is ably maintained in the Treatise on Harmony published by Dr. Holder, in 1694, a very clear and scientific work, in small compass.

(TTT), p. 198. See the Dict. de Musique of J. J. Rousseau (Art. Consonnance), a work characterized by all the vivacity and penetration of that writer's genius, and exhibiting an accurate and comprehensive knowledge of the subject of which it treats; though somewhat thrown into the shade by the unhappy celebrity of those pernicious productions with which his name is more commonly associated.

(YYY), p. 201. "Tartini considered the quintuple proportion as unfit for melody, and impossible to be executed. Time has shown that neither of these judgments is well founded."—Callcott's Grammar, p. 40. The instances, however, of this species of time seem to be extremely rare.

FINIS.

PRINTED BY J. MALLETT, WARDOUR STREET, LONDON.

Printed by Printforce, United Kingdom